The Decentralized Energy Revolution

The Decentralized Energy Revolution

Business Strategies for a New Paradigm

Christoph Burger and Jens Weinmann

© Christoph Burger & Jens Weinmann 2013

All rights reserved. No reproduction, copy or transmission of this publication may be made without written permission.

No portion of this publication may be reproduced, copied or transmitted save with written permission or in accordance with the provisions of the Copyright, Designs and Patents Act 1988, or under the terms of any licence permitting limited copying issued by the Copyright Licensing Agency, Saffron House, 6–10 Kirby Street, London EC1N 8TS.

Any person who does any unauthorized act in relation to this publication may be liable to criminal prosecution and civil claims for damages.

The authors have asserted their rights to be identified as the authors of this work in accordance with the Copyright, Designs and Patents Act 1988.

First published 2013 by
PALGRAVE MACMILLAN

Palgrave Macmillan in the UK is an imprint of Macmillan Publishers Limited, registered in England, company number 785998, of Houndmills, Basingstoke, Hampshire RG21 6XS.

Palgrave Macmillan in the US is a division of St Martin's Press LLC, 175 Fifth Avenue, New York, NY 10010.

Palgrave Macmillan is the global academic imprint of the above companies and has companies and representatives throughout the world.

Palgrave® and Macmillan® are registered trademarks in the United States, the United Kingdom, Europe and other countries.

ISBN: 978–1–137–27069–6

This book is printed on paper suitable for recycling and made from fully managed and sustained forest sources. Logging, pulping and manufacturing processes are expected to conform to the environmental regulations of the country of origin.

A catalogue record for this book is available from the British Library.

A catalog record for this book is available from the Library of Congress.

10 9 8 7 6 5 4 3 2 1
22 21 20 19 18 17 16 15 14 13

Printed and bound in Great Britain by
CPI Antony Rowe, Chippenham and Eastbourne

Cover image: Stylized detail of the US electricity distribution grid

Contents

List of Figures	ix
List of Boxes	x
Acknowledgments	xi
List of Abbreviations	xii

Introduction: Decentralized Energy as a Disruptive Innovation		1
The design of the future energy system		1
How to read this book		4
1	**Empowerment Paradigm – The Age of the Prosumer**	7
	Technological drivers	11
	Regulatory drivers	13
	Empowerment drivers	14
	Agents of change	15
2	**Small Is Beautiful**	17
	Theoretical framing	17
	Industry development	17
	Innovation competence of incumbents	18
	Micro CHP	19
	The efficiency advantage	19
	Exploiting scarcity pricing in the wholesale market: virtual power plants (LichtBlick)	32
	Manufacturing synergies and economies of scale (Volkswagen)	41
	Micro turbines	51
	Aviation technology for the energy market	51
	Market entry and sales strategies (Greenvironment)	54
	Cooperations and alliances (Greenvironment and Viessmann)	58
	Findings on micro CHPs and micro turbines	61
3	**The Rise of Island Systems**	64
	Theoretical framing	64
	Capabilities and collective empowerment	64
	Race to the top and strategic differentiation	66
	Bioenergy villages	68
	The countryside strikes back	68

Ownership and participatory processes (Jühnde)	69
Information dissemination and Smart Community consulting (Jühnde)	77
Recommunalization	80
The concept of citizen value (SWK, SWU, GASAG)	80
Findings on empowerment and recommunalization	84

4 Smart Management of Electricity and Information — 87

Theoretical framing	88
Network externalities	88
Standardization and lock-in effects	89
Smart grid	91
The feed-in induced revolution (Siemens, EnBW Regional, and ODR)	93
ICT to manage bidirectional power flows (Itron, Argentus, and ODR)	105
Smart meters	110
Competing standards (Itron)	110
Netting peak-shaving and increasing the share of flexible power demand (Itron, SWU, and Siemens)	119
Smart home and cross-selling opportunities (Siemens, E.ON)	125
Findings on smart management of electricity and information	127

5 Local Storage Solutions — 131

Stationary storage	131
The alternative to grid renewal	133
Developing the blueprint for carbon-free energy systems (Younicos)	136
Electric vehicles	140
A hype revisited	140
Combining lead markets and lead suppliers (Daimler)	144
Findings on stationary storage and electric vehicles	149

6 Enabling Negawatts — 150

Theoretical framing	150
Inelastic demand for durable goods	153
Split incentives and the principal-agent dilemma	155
Transaction costs and incomplete contracts	156
Building efficiency	158
Technological progress driven by standards?	158
Creating lasting ties with the building owner (co2online)	164
Energy performance contracting	169
From engineering to economics	169

	Product standardization as the key to customer management (Argentus)	178
	Risk reduction via guaranteed energy savings (Argentus)	181
	Findings on building efficiency and energy performance contracting	183

7 Insights from Germany for a Decentralized Energy Future — 185

1. The decentralization snowball — 185
2. Emotionalization of energy — 186
3. From single technologies to systemic viability — 187
4. Exploiting market volatility — 188
5. Guidance in the information tsunami — 189
6. From optimizing to satisficing — 189
7. Innovation and dissemination networks — 190
8. The right sequencing of the energy transformation — 191
9. From unbundling to rebundling — 192
10. Public service obligation for transparency — 193

Appendix: Company and Interviewee Profiles — 194

Argentus	194
Konrad Jerusalem	194
Bioenergy village Jühnde	195
Eckhard Fangmeier	195
co2online	196
Johannes Dietrich Hengstenberg	196
Daimler	197
Ulrich Müller	197
E.ON	198
Eckhardt Rümmler	198
EnBW Regional	199
Michael Kirsch	199
GASAG	200
Andreas Prohl	200
Greenvironment	201
Radu Anghel	201
Itron	202
Karsten Peterson	202
Werner Paech	202
LichtBlick	203
Ralph Kampwirth	203
ODR	204
Frank Hose	204
Siemens	205
Michael Weinhold	205

Stadtwerke Krefeld 206
Carsten Liedtke 206
Stadtwerke Unna 207
Christian Jänig 207
Viessmann 208
Manfred Greis 208
VW 209
Jürgen Willand 209
Younicos 209
Alexander Voigt 210

Notes 211

References 213

Key concepts, persons and technologies 219

Companies and organizations 221

List of Figures

1.1	Energy system trajectory	8
2.1	Efficiency gains of cogeneration, compared to conventional methods	20
2.2	The China Pavilion, an energy-saving building at the Expo 2010 in Shanghai	22
2.3	Residential fuel cell unit used in the ENE-Farm initiative	27
2.4	Global micro CHP unit sales by technology	28
2.5	Industrial production of micro CHP units at Volkswagen	46
2.6	Quiet Revolution turbine at Environment Energy Centre, Leyland	50
2.7	Micro turbine CHP plant with biogas in Muntscha	53
2.8	Recharging with the Nuru Power Cycle	62
3.1	Energy-supply concept of the bioenergy village Jühnde	68
3.2	Samsø's main energy production sites	70
3.3	Jühnde villagers in a meeting next to the biogas plant	76
3.4	Proposed master plan of Masdar City	79
3.5	Solar thermal heating in Dezhou	85
4.1	The externality problem with the design of smart grids	89
4.2	Components of the smart grid	92
4.3	Ownership of decentralized energy generation units in Germany	100
4.4	Smart metering solution by Siemens	113
4.5	Smart home vision by General Electric	125
6.1	McKinsey abatement-cost curve for the buildings sector in Germany, year 2020	152
6.2	Share of German heating systems replaced annually	154
6.3	Energy efficiency measures implemented in commercial buildings	158
6.4	Celebration at Druk White Lotus School, India	170
6.5	The concept of energy performance contracting	172

List of Boxes

2.1	Heat and Power Cogeneration at the World Expo 2010, Shanghai	22
2.2	"ENE FARM" – commercialization of a residential micro CHP unit based on fuel cells	27
2.3	Small wind turbines – innovations for downsizing in urban settings	50
2.4	Energy poverty, village level entrepreneurs, and the Nuru Power Cycle	62
3.1	Samsø – energy autonomy in the Baltic Sea	70
3.2	Masdar – island solutions under harsh environmental conditions	79
3.3	Dezhou and the Chinese Solar Valley – reconciling growth with environmental protection	85
4.1	A smart grid for a Smart City – Boulder paves the way and shows the obstacles	105
4.2	Telegestore in Italy – benefits of a mass rollout of smart meters	128
6.1	Building efficiency in developing countries – Druk White Lotus School in Ladakh, India	170
6.2	Weather forecasts to increase the energy efficiency of intelligent buildings	182

Acknowledgments

First and foremost, we would like to express our gratitude to the interviewees, who agreed to share with us their experiences and spent a considerable amount of time revising and improving their individual narratives. Their enthusiasm for decentralized energy gave us the inspiration and endurance to write this book.

We are also indebted to Jinfeng Ding and Chen Huihuang from the China Executive Leadership Academy Pudong (CELAP) for providing information on our Shanghai business case.

We wish to thank our colleagues at the European School of Management and Technology for their intellectual, academic, and logistic support during the writing phase, in particular Derek Abell, Olaf Plötner, Mario Rese, and Gabriele Weber, as well as Bianca Schmitz and Frauke Wriedt. Last, but not least, we would like to extend special thanks to Eleanor Davey-Corrigan at Palgrave Macmillan and Vidhya Jayaprakash at Newgen Knowledge Works.

List of Abbreviations

CCGT	combined-cycle gas turbine
CHP	combined heat and power
CO_2	carbon dioxide
EEG	Erneuerbare Energien Gesetz (Renewable Energies Act)
GW	gigawatt
ICT	information and communication technology
IEA	International Energy Agency
IFMA	International Facility Management Association
kW_{el}	kilowatt
kWh	kilowatt hour
kW_{th}	kilowatt (thermal)
MW	megawatt
MWh	megawatt hour
SME	small and medium enterprises
SWK	Stadtwerke Krefeld (municipal utility Krefeld)
SWU	Stadtwerke Unna (municipal utility Unna)
VW	Volkswagen

Introduction: Decentralized Energy as a Disruptive Innovation

The design of the future energy system

Disruptive innovations threaten the market dominance of established producers of goods and services, as Harvard professor Clayton Christensen describes it. The producers of these innovations frequently start off with low gross margins and a small target market, but they incrementally move up the scale and may eventually displace established competitors. Their success strategy is to offer a different value proposition than the one being offered by incumbent players; even if the products are inferior in some aspects, consumers are willing to buy in. Many industries have been shattered and reorganized after disruptive innovations, including the computer industry (from mainframes to laptops and tablets), telecommunications (from landlines to cell phones and smart phones), and camera manufacturers (from analog to digital).

The energy supply industry already underwent two significant changes in the recent past: First, liberalization imposed the unbundling of vertically integrated utilities and the introduction of competition in the power generation and retail business; second, subsidies for energy generation from renewable sources led to a massive increase in wind farms and photovoltaic panels in many countries. Those two changes are small compared to the disruption that a decentralized energy supply could create for the energy system, its industrial structure, and the role of the consumer.

The value proposition offered by decentralized generation differs fundamentally from the current energy system configuration: It turns the one-way street from producer to consumer upside-down. It enables every household, as well as all types of commercial and industrial consumers, to become active agents and autonomous providers of energy, either for own consumption purposes or to generate revenues by feeding electricity into the central grid.

A decentralized energy supply satisfies two of the three dimensions of the so-called energy policy triangle – environmental sustainability and security of supply. It is environmentally friendly because most of the technologies used in a decentralized energy supply are based on renewable energies or highly efficient conventional technologies such as cogeneration that serve as a bridge toward a carbon-free supply system. Decentralized energy also increases the security of supply because it harvests local resources and decreases dependence on finite energy reserves. However, the third dimension of the energy policy triangle – cost-effectiveness – is a major obstacle in the deployment of decentralized energy sources: many of the devices are in an early stage of technological maturity, for example micro cogeneration plants or stationary batteries; or they occupy niche positions and are not yet produced in larger numbers, like micro turbines; or they are simply too expensive or inconvenient to be accepted by final consumers, as is currently the case with smart meters.

The industry structure of the energy sector does not provide the best conditions for new technologies to emerge and spread. Apart from hesitant incumbents who may stick to centralized generation as their established business model, the systemic features of energy supply prevent innovations from being implemented overnight. As a large technical system, it exhibits substantial inertia due to technical standards, organizational routines, and a highly capital-intensive infrastructure.

While most reference scenarios for energy supply predict that renewable energies and decentralized technologies will capture a high share of the energy mix by mid-century, the question arises whether decentralized energy supply has the potential to become a disruptive innovation and initiate a global paradigm shift in the near future.

Even if there is consensus among experts that the economic potential for decentralized energy supply is substantial, not much literature exists that explores *how* to get there: More specifically, which business strategies are established by entrepreneurs working in the field of decentralized energy supply? How do they overcome economic, technical, and psychological hurdles – and, in particular, do they manage to set up business models that survive without subsidies? What responses and adaptation tactics are chosen by incumbent utilities to cope with the dilemma of maintaining their competitive advantage and leading positions while building up competences for a new and potentially disruptive market in decentralized energy solutions? Who will be the new entrants as the boundaries of the energy sector gradually dissolve? Diversifiers from the information and telecommunications industry, mechanical and electronic engineering firms, and even car manufacturers are bringing fresh ideas into the sector. Which roles will they take?

This book contains the findings and extended narratives of a series of 17 semi-structured interviews with decision-makers working toward a decentralized energy supply. Entrepreneurs, incumbents, and community leaders provide first-hand insights on the strategies they pursue; the commercial, technical, and regulatory challenges ahead; and the prospects for a decentralized energy system to overcome the current status quo. We integrated energy and non-energy firms into our sample and blended their individual narratives with comments and contrasting viewpoints from other players.[1]

We communicated with our interview partners during the period between 2011 and mid-2012, when we finalized the publication. The nuclear accident in the Japanese power plant Fukushima occurred during this period. It may have an impact on the speed of change, but none of our interview partners expressed that the accident had fundamentally altered the strategic stance towards decentralized energy supply.

Energy regulation exists within a highly dynamic regulatory context; therefore we tried to extend the expiry date of our insights by omitting comments too closely linked to specific regulations and systemic features that may be adapted or overthrown in the near future. We rather focus on business strategies that are likely to be relevant until decentralized energy supply leaves its niche status and becomes a central feature of our energy world.

Even though many of the companies discussed in this book operate in multiple countries, the interviews concentrate on activities in Germany. The German government has decided to radically transform its energy system. The objective is to become both carbon-free and nuclear-free over the coming decades. For all industrialized countries in the world where the existing energy infrastructure is well-established and reliable, Germany is likely to serve as

Table I.1 List of firms of the interviewees

	Incumbents	Municipal utilities	New entrants
Energy	*Energy and gas utilities* E.ON GASAG	*Integrated utilities*	*Micro turbines* Greenvironment
	Regional grid operators EnBW Regional ODR		*Virtual power plants* LichtBlick
	Manufacturers Siemens Viessmann	SWK SWU	*Island solutions and storage* Younicos
	Smart meters Itron		*Building efficiency* co2online
	Electric vehicles Daimler		*Energy performance contracting* Argentus Energie
Non-energy	*Micro cogeneration* VW		*Bioenergy villages* Jühnde

a role model for the transformation of the energy sector and the emergence and success of a decentralized energy supply.[2] Germany has a successful track record of environmental regulation – the green movement had already begun influencing politics in the 1970s; it has a long history of entrepreneurial activity and organizational diversity of its small and medium-sized enterprises; and it has a federal structure that allows for experimentation and environmental protection on all political levels. "The future energy system will be designed in Germany," one of the interviewees stated.

How to read this book

The book focuses on technologies and systemic solutions in decentralized energy generation that are likely to leave their niche status but have not yet reached mass scale, including:

- micro combined heat and power (CHP) plants and micro gas turbines (Chapter 2);
- bioenergy villages and communal projects (Chapter 3);
- smart grids and smart meters (Chapter 4);
- stationary and vehicle-based storage (Chapter 5);
- building efficiency and energy service contracting (Chapter 6).

We start with a brief review of the main drivers of the decentralized energy revolution in Chapter 1. While Chapter 2 explores functional niches and intelligent micro supply, Chapter 3 focuses on regional niches with island systems and recommunalization. In Chapters 4 and 5, we analyze how a decentralized energy supply will affect the electricity grid. We titled Chapter 6 "Enabling Negawatts" because we believe that building efficiency and energy performance contracting are integral parts of decentralized energy solutions. They typically do not add generation to the system but rather reduce the overall consumption and make energy use more efficient. On the level of the final consumer, they are as decentralized as can be. We conclude with ten insights from Germany in Chapter 7.

Efficiency improvements in industry and the manufacturing sector offer a massive savings potential, too, but a thorough discussion would exceed the scope of this book. Similarly, photovoltaic panels and wind farms are not explicitly discussed, mainly because they have already left their niche status. They, of course, play a role in many of the topics discussed in the book – in particular the smart grid and storage solutions – and integrated business strategies will still be in demand after subsidies have ceased.

Although we primarily focus on the electricity system, we decided to title the book *The Decentralized Energy Revolution* because calorific energy plays an important role in combined heat and power. Moreover, building efficiency and energy performance contracting also include strategies to economically optimize heating and cooling requirements of houses.

Some of the business strategies are closely linked to market failures or market barriers, and how to overcome them. Wherever the academic literature in the fields of economic theory, strategy, and innovation provides insights as to why obstacles exist or how to frame and structure a market feature, we introduce them at the beginning of the respective chapter.

Qualitative research sometimes tends to suffer from a positive selection bias – all business strategies presented here are somewhat successful examples of the creation of new organizations, the reorientation of existing businesses, or the emergence of ambidextrous structures within larger firms. Outright business failures – which would almost naturally complement this sample, given the process of creative destruction apparent in all system transformations – are exempted. However, this book's success stories also entail accounts of initial failures, paths that were not pursued, and periods of crisis and reorientation that founders had to undergo and overcome.

In many parts of the world, entrepreneurs come up with locally adapted, successful decentralized solutions, governments launch new and promising policy incentives, and companies implement innovative pilot projects. Throughout the book, the insights from Germany will be complemented with case studies from a broad range of countries – including other European countries, China, India, Japan, countries in Sub-Saharan Africa, the United States, and the United Arab Emirates – to exemplify the strategic diversity and creativeness inherent to any process of change.

Some readers may be particularly interested in a specific technology and its prospects; others may find it most revealing to discover patterns on how entrepreneurs overcome hurdles and market barriers in order to implement innovative strategies. A third group of readers could be attracted to comparing differing, often controversial views of individual agents or companies on specific topics. To satisfy these preferences and reading habits, we decided to implement a genuinely modular structure, such that each chapter and each strategy can be read and understood on a stand-alone basis. In addition, the company index at the end of the book indicates all pages on which a representative of a specific company is quoted. The Appendix provides short biographical notes on the interviewees and an overview of the history and current status of their respective firms.

This book focuses on business strategies for a decentralized energy system. It tackles technological challenges only marginally. For a more in-depth discussion of the engineering part, we recommend specialized books on decentralized cogeneration (Chamra and Mago, 2009, Kolanowski, 2008, Pehnt et al., 2010, Borbely and Kreider, 2001) and smart networks (Galvin and

Yeager, 2008, Kaplan and Net, 2009, Bauknecht, 2012). For more general-interest books about decentralized features of the future energy system and how society has to change to cope with climate change, interested readers may want to direct their attention toward *Powering the Future* by Nobel Prize winner Robert B. Laughlin (2011) and *The Citizen-Powered Energy Handbook: Community Solutions to a Global Crisis* (2007) by Greg Pahl. Certainly the most influential of all recent books on decentralized energy supply, Jeremy Rifkin's *The Third Industrial Revolution* (2011), has already motivated the European Parliament to issue a formal declaration calling for the implementation of a new phase in energy policy, and we would recommend to complement our endeavor with Rifkin's analysis.

1
Empowerment Paradigm – The Age of the Prosumer

The history of humankind is inextricably intertwined with the usage of energy. As much as human development was characterized by slow adaptation processes and sudden social or cultural revolutions, the shifts in energy use can also be interpreted as periods of slowly evolving, incremental progress and abrupt – and often radical – changes. Since the very beginning of cultural organization, human beings used manual force, mechanical tools, and the power of domesticated animals to make their labors easier and create an ever-increasing quality of life. While the taming of fire facilitated the use of fuel wood, peat, and dung for cooking and heating, very early civilizations already knew how to extract energy from water and wind. The sun served for drying food and producing salt.

Biomass – and later also charcoal produced from biomass – remained the single most important source of energy for most of our history. With the demographic growth from the 16th and 17th centuries onwards, the depletion of forestry reserves and the rise of the steam engine during the industrial revolution required – in particular in Europe – a reorientation in the use of energy sources; coal – with a much higher energy output per unit mass than wood or charcoal – incrementally replaced timber as the most important energy source for human beings, reaching a first peak in the first half of the 20th century. Meanwhile, oil extraction became technically feasible, and with the rise of individual vehicle transport, oil replaced coal as the predominant form of energy after the 1950s. The usage of natural gas, albeit technically more demanding because of its grid character, also began on a larger scale in the second half of the past century. While hydroelectric power has had a minor but not unimportant role in electricity generation since the late 19th century, the emergence and usage of nuclear fission proved to be the last major new entry among the top primary energy sources.

Today, a new transformation is about to take place: Climate change and the threat of global warming, as well as finite fossil resources, require that societies step back from carbon dioxide-intensive forms of energy (for a thorough discussion, see Gore, 2009). A growing public skepticism toward nuclear power prevents this largely carbon-free technology to substitute fossil fuels on a large scale. Humankind seems to be on the way to returning to a larger share of renewable, non-polluting sources of energy.

The cyclicality in the energy supply mix is paralleled by similar developments in the balance between central and decentralized energy generation. While early civilizations collected fuel wood and produced charcoal in a local setting, the industrial revolution led to larger power plants for manufacturing and electricity provision. In the 20th century, economies of scale reached their peak with the introduction of nuclear power plants and large hydropower dams. But with liberalization, smaller combined cycle gas power plants became attractive because of their standardized design, lower investment costs, and high efficiency. The move to small-scale supply structures was accelerated by advances in photovoltaic panels, solar thermal heating, and wind turbines.

Meanwhile, the ownership structure of assets in the energy sector shows the same pattern: From dispersed ownership of the first energy-harvesting installations to publicly owned utilities in many countries during the 20th century, and back to private investors after liberalization. In Germany, more than half of the capacity of renewable energies is owned by private persons and farmers. Figure 1.1 shows how the energy system trajectory may eventually return to its starting point with a decentralized, carbon-free and privately owned energy supply system.

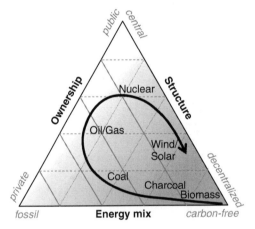

Figure 1.1 Energy system trajectory

However, renewable energies have their drawbacks. First, as compared to fossil fuels, their energy yields are often low, and harvesting them requires sophisticated, expensive technologies. Second, their availability typically fluctuates over time and varies according to geography and climate. The dispersed settings pose a major challenge because the energy supply structure has to move back from a system with power plants based on fossil resources that start operating upon request to a structure that absorbs energy where and when it becomes available, often without the convenience of being able to store larger amounts of energy over time.

While early human civilizations consumed energy in modest amounts – both per capita and in total – the fundamental challenge of the return to decentralized, renewable energy supply is how to satisfy the energy needs of 7 billion people or more without sacrificing the generally high quality of life of people in industrialized nations or without jeopardizing the prospect of reaching the same quality of life in developing and emerging countries.

This challenge requires a fundamental, far-reaching change in the supply patterns and can be seen as a paradigm shift. Following Markard and Truffer (2006, 611), we define paradigm, in this context, as a "prevailing model for the solution of techno-economic problems." Paradigm shifts in large technical systems occur less frequently than in other fields of industrial activity because the technical interdependencies of system components, their standards, institutions, and routines create a high degree of path-dependency on the overall configuration of the system, in particular in grid-based energy services.

In the electricity sector, a sequence of three paradigms can be postulated, which we call "engineering paradigm" for the early stage of energy services, "economics paradigm" for the epoch since the liberalization of the sector, and "empowerment paradigm" for the decentralized energy supply of the future.

The first phase – the engineering paradigm – was characterized by competing technologies closely linked to individual entrepreneurs. Thomas Alva Edison developed a long-lasting light bulb in 1879 and established the first direct current (DC) electric power distribution system in 1882, which supplied street lamps and private houses with electricity – in total 59 customers. His business model spread across cities in the United States within a decade, and he formed the Edison General Electric Company. However, one of the main disadvantages of the DC technology was that the current could not be transported farther than around three miles without major losses, so there were decoupled, independent patches with autonomous electricity provision.

Meanwhile, another US entrepreneur, George Westinghouse, developed a commercial application of Nikola Tesla's alternating current (AC) transmission system and installed the first commercial AC power system in 1891 in the US mountain resort of Telluride, Colorado. Westinghouse and Edison entered the so-called War of Currents, in which each propagated his system as being superior. Edison pointed to the dangers of alternating current and campaigned against his competitor by electrocuting animals, including the by-then famous Topsy the Elephant. Ultimately, Edison's efforts did not pay off. In 1893, Westinghouse won the competition to provide electricity for the World's Fair in Chicago and was selected to construct the generators for the hydroelectric power plant at Niagara Falls, USA. The AC technology eventually became the US standard. At the turn of the century, electricity had reached all industrialized countries and drastically altered public and private life, transportation, and industry.

After the First World War, governments started to strive for standardization and control of the electricity system, with the aim of establishing integrated, large-scale grid supply structures within their respective countries. By the late 1930s, much of the formerly privately owned and locally organized firms were absorbed by the overarching, domestic transmission networks, most often vertically integrated and with a strict tariff control based on rate-of-return regulation. With the emergence of electric devices like vacuum cleaners, refrigerators, and washing machines, electricity moved from being an integrated service like lighting to a basic infrastructure requirement for multiple appliances.

The electricity system under the engineering paradigm proved to be fairly stable and reliable. However, a number of influential scholars, in particular Noble prize winner Milton Friedman and his peers at the Chicago School of Economics, heralded the economics paradigm, when they expressed general skepticism about too heavy involvement of the state in the economy. Their view spilled over into the Western political elite in the 1980s. US president Ronald Reagan and UK prime minister Margaret Thatcher started to introduce competition in state-controlled infrastructure services, state assets were privatized, and financial institutions such as the World Bank and the International Monetary Fund coupled free-market policy directives with their structural adjustment loans for developing countries (Stiglitz, 2002). In continental Europe, the European Commission became a key proponent of liberalization because the concept closely matched the vision of a single European market across all member states.

Electricity became a commodity. Liberalization reallocated ownership, dismantled vertically integrated utilities, reshuffled control rights, and introduced market elements such as wholesale spot trading and retail competition. A range of new market players entered the sector with innovative strategies, such as Roger W. Sant and Dennis W. Bakke, the founders of AES, which became one of the first firms to see and exploit the full potential of liberalized markets in the United States. Meanwhile, incumbent utilities were deprived of secure revenue streams and had to reconfigure their traditional business models.

The liberalization of grid-based services was a top-down regulatory change, which also had consequences for the technologies deployed. In particular, gas-fired, combined-cycle turbines became the preferred choice for investments in generation capacity (Markard and Truffer, 2006, 616). Due to the economics of series production, a high degree of standardization in the construction of power plants – and their economic and energetic efficiency – they altered the power plant mix of many countries, for example the United Kingdom and the United States. Simultaneously, advances in information and communication technologies (ICTs) allowed for the participation of many industrial and commercial players in wholesale and retail markets.

Public funding of renewable energies, in particular wind and solar power, triggered the beginning of the empowerment paradigm. The subsidies succeeded in attracting a multitude of private investors – often on the community and even household levels – to install photovoltaic panels or wind parks. This had major consequences for the transition to a sustainable energy supply structure, but also created substantial challenges to grid stability and the coordination of loads and energy flows. Closely connected to the deployment of renewable energies, decentralized energy solutions beyond wind and solar emerge as parallel features of the energy system. Compared to the top-down changes of liberalization and renewable energy subsidy schemes, the move toward a decentralized energy supply is more complex in its forces, as we will outline in the next sections.

Once a new paradigm has been established, a period of upheaval is followed by consolidation and continuity. Institutions are created that provide the administrative framework, ensure legal compliance, and maintain the status quo. While the early phase of electrification was characterized by entrepreneurial activities, in the early 20th century national governments took over and defined electricity supply as a core infrastructure and part of the public service obligation that the state had to take care of. The transition to the economics paradigm was accompanied by legislative changes such as the Purpa Act in the United States or the Lisbon Treaty in the EU. The responsibilities for regulating and monitoring the market once it was established were taken over by institutions like the European Commission and national regulatory agencies with formal independence from their respective governments. In the EU, the regulators have since then coordinated and advanced the market design under the existing status quo.

Table 1.1, on the following page, sketches the sequence of three paradigms with its main agents and institutions.

We have identified three key drivers for the transition to the new paradigm: Technologies, regulation, and empowerment.

Technological drivers

A decentralized energy supply relies on a multitude of individual technologies. Some of them have already been on the market for several decades but have not left their niches, such as heat pumps and micro CHP units. It is only now that industrial mass production has created the necessary economies of scale to reduce investment costs. As one example, we will present the procedures that Volkswagen has introduced to turn the manufacturing of micro CHP units into a venture that utilizes the mechanisms and routines that the company

Table 1.1 Paradigm shifts in the electricity sector

	Crisis/trigger of change	System construction and consolidation
Engineering paradigm **Past:** Phase of emergence and consolidation of centralized energy system structure	**Economic:** – Increasing wealth and the desire for amenities of modern civilization **Technological** – Invention of the electric motor and generator **Scientific** – The discovery of electromagnetic forces *Agents of change* Edison Westinghouse	– Provision of public service as state responsibility – Vertically integrated, state-controlled utilities – Economies of scale *Agents of continuity* National governments Energy incumbents
Economics paradigm **Present:** Phase of liberalization and globalization	**Economic:** – Inefficiency of central system – State: budget deficits – Capital markets able to finance private projects **Technological** – Mid-scale becomes affordable (dis-economies of scale: CCGT) – ICT to coordinate industrial and commercial agents in markets **Political** – Neoliberal ideology and pro-competitive policy agenda – EU: Single European market **Scientific** – Neoliberal economic theory *Agents of change* Reagan Thatcher Friedman (Chicago School) Sant and Bakke (AES Founders)	– Several regulatory models in parallel, but most often unbundling of generation, transmission and distribution – Competition in generation and retail, transmission and distribution cost regulation – Supranational integration *Agents of continuity* European Commission National regulatory authorities like FERC, Ofgem, or BNetzA

Empowerment paradigm
 	
Future: Phase of decentralization and decarbonization
	Economic:
	– Internalization of externalities (esp. CO_2)
	– Reduction of import-dependence
	– Finite fossil resources
	Technological
	– Micro scale of generation technologies (CHP, turbines, meters)
	– ICT to coordinate residential agents in markets
	Political
	– Recommunalization
	– Environmental awareness
	
	Agents of change
	Hajee (see p. 61)
	Voigt (Founder of Younicos)
	Hermansen (see p. 69)
		– Consumers as producers
		– Parallel systems:
		  • Central / decentralized generation
		  • Multi-country transmission grids / island grids
		
		Agents of continuity
		CEN (standardization)
		CEER & ERGEG (coordinated regulation)

employs to produce cars. Similarly, batteries for power storage have experienced steep cost degressions over the past couple of years, and will become much less expensive in the near future because improvements in battery technology for electric vehicles will create significant spillover effects, especially to stationary applications.

Most importantly, information and communication technologies have progressed at an impressive pace over the past decade. The decentralized energy revolution would not be possible without the digital revolution. ICTs allow for the instantaneous coordination of multiple agents in a large technical system. The current grid structure will have a "smart" layer added on top that will transmit all system states and provide the basis for largely autonomous island systems.

The combination of new energy technologies with new ways of communicating may even trigger a "third industrial revolution," as Jeremy Rifkin suggests:

> In my explorations, I came to realize that the great economic revolutions in history occur when new communication technologies converge with new energy systems. New energy regimes make possible the creation of more interdependent economic activity and expanded commercial exchange as well as facilitate more dense and inclusive social relationships. The accompanying communication revolutions become the means to organize and manage the new temporal and spatial dynamics that arise from new energy systems. (2011, 2)

The parallel development of new communication and energy technologies is a necessary but not sufficient condition for initiating a paradigm shift, though. It requires the market, hence the consumers, to accept and buy new products. Smart meters or electric vehicles are prime examples that new technologies are available but that they may not be adopted unless regulatory incentives are introduced.

Regulatory drivers

Energy supply is part of the core infrastructure services, for which the government bears ultimate responsibility. Even in liberalized markets, a total withdrawal of the state from the sector is practically impossible. The foresight of political decision-makers and regulatory agencies is supposed to correct for market failures, in particular negative externalities due to climate change and security of supply.

Decentralized energy generation has the advantage that many of the technologies are less carbon-intensive and more efficient than conventional energy production. In particular, the efficiency of cogeneration, that is, the coordinated production of heat and electricity in a single plant, exceeds the production of heat and power in individual plants by far. Similarly, increasing the efficiency of the housing stock often provides a least-cost option to save tons of carbon dioxide while reducing the energy bill of homeowners.

In addition to measures that mitigate climate change, decentralized energy generation increases energy autarky. The use of local biomass, wind power, or geothermal energy, including heat pumps, reduces the requirement to import fossil fuels. Given the finite amount of all fossil reserves and growth in global demand, rising prices are likely to be observed in the future. Domestic renewable energy sources provide an effective means to hedge against international price fluctuations and are therefore politically welcomed.

Regulation to promote decentralized energy generation turns out to be very successful in many countries. Residential users installed their private generation devices along the local distribution grid, often with a capacity that far exceeds their own loads and allows for extensive feed-in. Regulatory agencies have to find and impose adequate mechanisms to secure the availability of supply to match inelastic demand.

Ultimately, decentralized energy generation may save the taxpayers' money. The construction of costly electricity transmission lines and a reinforcement of the local distribution grid can be avoided if regulators promote a smart grid with a localized balancing of demand and supply and if, for example, small cogeneration plants or stationary storage devices provide the services necessary to maintain local or overall system stability.

Empowerment drivers

Globalization and free trade have not only brought increasing wealth and an unprecedented amount of goods and services to consumers; they are also often blamed for financial turmoil, economic instability, and rapid environmental degradation. As an immediate response, ecologically aware consumers prefer produce of regional origin – "think global, act local" is a slogan that found supporters in many countries. The UN climate conference in Rio de Janeiro in 1992 initiated and institutionalized the Local Agenda 21, which promoted the self-organization of communities to define their individual targets for better living and sustainability.

Even though liberalization of infrastructure services was heavily criticized as being part of the neo-liberal agenda of the proponents of globalization, it set the stage for an increase in consumer choice, in particular in grid-based energy services, telecommunications, and transport. Whereas previously, for example, the electricity provider was pre-determined by territorial boundaries of the utility, competition in retail markets would now enable residential consumers in many countries to freely choose an electricity retailer that reflected their preferences, for example with carbon-free electricity.

Meanwhile, modern societies evolve from hierarchies to networks (Mayntz, 1993), and new actor configurations give more power to the consumer than ever before. Grassroots movements have benefitted from the rise of the digital society; best (and worst) practices can now be communicated instantaneously to a cross-continental audience. Community learning occurs on a global scale.

Decentralized energy generation parallels the rise of the internet, which turned passive consumers of television and radio into active and interactive contributors to the worldwide web. It transforms the consumer into a "prosumer," who produces energy and feeds it into the overall system. This happens both on the individual level – when residents decide to install a micro CHP unit in their basements – and on the community level, when villages switch to a heat and power supply based on local biomass. Like many initiatives on the internet, say Wikipedia or the numerous blogs, wealth maximization and cost-savings often play a minor role in that decision process – more important is the power that is given back to the consumer in a field that has long been dominated by central planning and command-and-control.

Agents of change

A revolution requires courageous individuals who dare to challenge the existing paradigm and act as triggers to overhaul the status quo. Edison and Westinghouse initiated the rise of the engineering paradigm; Thatcher and Reagan pushed the world to introduce the economics paradigm. Decentralized energy generation and the empowerment paradigm also have key proponents among the global political and industrial elite – such as Klaus Töpfer, the founding director of the Institute for Advanced Sustainability Studies (IASS) and former Executive Director of the United Nations Environment Programme.[1]

But the empowerment paradigm is fueled by a multitude of entrepreneurs who develop appropriate and locally adapted small-scale solutions. In Table 1.1 we have mentioned just three of them: Sameer Hajee, who was nominated "Social Entrepreneur of the Year in Africa" by the Schwab Foundation for Social Entrepreneurship in 2012 (see Box 2.4 on p. 61); Søren Hermansen, a high school teacher who convinced Danish islanders to abandon their fossil-based energy supply and become auto-subsistent with renewable energies (see Box 3.1 on p. 69); and Alexander Voigt, the founder of island solutions provider Younicos, whose business model will be discussed in detail in Chapter 3.

These and many other agents of change will shatter the established structure of the energy industry. "Novelty emerges only with difficulty, manifested by resistance, against a background provided by expectation," Thomas Kuhn wrote about scientific revolutions (1962, 64), but these insights may also be applied to the energy sector. Alexander Voigt comments:

> Every revolution in market systems has led to the giants disappearing or having their power slashed. In the telecommunications sector, for example, small firms with new ideas emerged and grew very fast. When we liberalized the telecommunications market, who would have reckoned with Google, Skype, or Twitter? These companies grew from a base as start-ups. And that is how we have to look at the energy market. Supplying services and new products in the energy market will certainly become more the domain of newcomers and less of those firms that are still stuck in traditional models. They are simply not capable of it. If they were, they would do it.
>
> There will continue to be asset owners sitting on assets worth billions. But what role will they play in value creation and what margins can be achieved? That is the decisive point with respect to competitiveness. The big energy utilities are certainly developing a new awareness but much too slowly. The energy market will not be dominated by those companies that own wind parks in the North Sea.
>
> If the politicians shape the market in the right way, start-ups and new players will enter the market, putting the big players under extreme pressure. Big companies from related areas will enter in the energy sector. Telecommunications enterprises, for example, with their competence in the IT area, in creating structures and in billing, have no problems offering electricity products as well. In a market in which the sale of electricity is separated from generation and in which there are wholesale markets, some of them will be very successful. (Voigt, Younicos)

In the following four chapters, we will explore how agents of change – newcomers but also pioneers and strategists from inside larger corporations – are implementing their strategies for a decentralized energy supply.

2
Small Is Beautiful

Over the course of the 20th century, economies of scale led to an ever-increasing size of power plants in the electricity sector. In the 1980s, technological progress and a shift in energy policy brought a reversal of that trend, though: combined-cycle gas turbines (CCGT) emerged, and entrepreneurial companies like AES in the United States used the emergence and regulatory implementation of the economics paradigm to establish themselves as new players in independent power production, often with small-scale, standardized natural gas plants, which pose substantially less investment risks than coal or nuclear power plants.

While some technologies still try to seize economies of scale – one example would be the repowering of wind turbines – other technologies are following the downward trend and are becoming increasingly miniaturized. How far can the downscaling go? When do decentralized supply installations become economically unviable and technically infeasible?

This chapter analyzes business models in the fields of micro cogeneration plants and micro turbines. These two underlying technologies have a long history in larger-scale generation, but they offer chances to become central pillars of a decentralized energy supply in the near future. In addition, several boxes depict successful practices of downsizing in fuel cell technology, Stirling engines, and wind turbines.

Theoretical framing

Industry development

Markets are not static. They rather develop and change their characteristics over time. Consulting practice Arthur D. Little was one of the first companies that developed a so-called product or industry life cycle (Arthur D. Little, 1981).

A market typically starts in an embryonic phase in which a basic product exists and a business model is developed. With increasing growth rates, new competitors enter the market as there are low barriers to entry. When the business model is established by suppliers and customer acceptance is achieved, cost as a competitive differentiator becomes increasingly important, leading to economies of scale as an underlying theme of this growth phase. In the maturity phase, the growth rate of the industry is below that of GDP but still remains positive, while in the ageing phase the growth of the markets starts decreasing.

During the industry phases, the role of technology changes. While in the embryonic phase, concept development and product engineering are the focus, this changes to product line refinement and extension (growth phase), evolution of processes and product line renewal (maturity phase), to process development and cost reduction (aging phase).

Characteristic company strategies change, too. At the beginning of the market development, taking risk, developing technology, rapid response, and gaining market share are key components. During the growth phase, industrialization in manufacturing, distribution, and service is a dominant theme, while rationalization and managing investment are characteristic strategies in the maturity and aging phases.

Deans *et al.* (2002) applied the life cycle from the perspective of consolidation, and mergers and acquisitions. They distinguish between the following: an opening phase with no or little market concentration; a scaling phase, where dominant players develop and a maximum consolidation speed is achieved; a focus phase, where the top three players increase their market share from 45 percent (scale phase) to about 70–80 percent; and a balance/alliance phase, where acquisitions and takeovers are difficult to realize and alliances offset mergers. In their research, the global consolidation process lasts 25 years at a standard deviation of five years and with a trend toward shorter cycles. They also note that the success rate of mergers and acquisitions over the phases decreases from 59 percent in the opening phase to 58 percent in the scaling phase, to 50 percent in the focus phase, and to 30 percent in the balance/alliance phase.

Innovation competence of incumbents

Schumpeter (1934) established the notion of creative destruction, wherein small, new players and entrepreneurs challenge current business practices pursued by incumbents of the industry.

In this environment, incumbents are faced with the dilemma to either participate in the new markets or fight against innovations to preserve existing business models. The key question for them is how they can benefit from new markets and innovations while remaining successful in their traditional businesses.

Several explanations have been offered to throw light on the lack of competence of incumbents to drive disruptive innovations. Economists argue that incentives differ between incumbents and new entrants. As incumbents aim to defend their market power, they pursue incremental – and not disruptive – innovations (Henderson, 1993). Organizational theory suggests that inertia constrains incumbents' efforts to become innovative (Hannan and Freeman, 1984). Strategy theory proponents argue that a company is embedded in a value network of suppliers, customers, competitors, investors, communities, etc., in which they have made strategic commitments (Christensen, 1997, Christensen and Rosenbloom, 2000, Ghemawat, 1991, Sull et al., 1997).

Yet, some incumbents have managed through periods of disruptive innovation, and even industries such as the pharmaceutical industry seem to survive and prosper (Rothaermel, 2001). Hill and Rothaermel (2003) propose several explanations for this success: Basic research can help to raise awareness of emergent technologies and enable the company to accumulate the necessary capabilities; a real-option perspective for evaluating technologies can help to overcome under-valuing investment in disruptive innovations; the legitimization and institutionalization of autonomous action can fight inertia; successful changes with regard to disruptive technologies in the past can create the basis for future changes facing new disruptive innovations; and finally, accumulated slack can be the resource for new efforts.

O'Reilly III and Tushman (2004) analyzed whether incumbents can remain successful in their traditional business model while also participating in disruptive innovations. They found that organizational structure and management practices play a key role in whether those efforts are successful or not. In their research, they differentiated four organizational designs to innovate: functional designs that integrate project teams in the existing organizational and management structure; unsupported teams that are set up outside the existing organization and management structure; cross-functional teams that operate within the organizational structure but outside the existing management hierarchy; and ambidextrous organization in which project teams are structurally independent units that have their own processes, structures, and cultures, but are integrated in the existing management hierarchy. In examining 35 attempts to participate in disruptive innovations while still remaining successful in the traditional business, they found that more than 90 percent of those using the ambidextrous organization succeeded. Only 25 percent of the functional designs achieved their goals, while cross-functional and unsupported teams failed completely.

Micro CHP

The efficiency advantage

Combined heat and power, also called cogeneration, is one of the most efficient ways to use primary energy. The technology allows for various primary energy intakes, including conventional diesel or even coal engines. Natural gas, however, has been the preferred fossil fuel over the past years. Cogeneration also operates with renewable sources, especially biomass, or industrial gases available for combustion. Cogeneration units typically consist of four elements:

a prime mover (engine or drive system), an electricity generator, a heat-recovery system, and a control system (IEA, 2008, 10).

The basic efficiency advantages of CHP emerge from the optimization between the high-quality electricity output and the lower-quality heat output. While pure electricity-generating units that are based on combustion technologies lose a great share of their low-quality heat in cooling towers or waste heat in rivers, cogeneration recuperates a substantial fraction of that heat for thermal usage, while only slightly reducing the power output of the unit. Compared to separate heat and power generation, a CHP unit can thus reach substantially higher energy efficiency.

Figure 2.1 exemplifies the basic principle of cogeneration: For the same heat and power output of CHP and conventional methods, that is, separate production of heat and power (in total 260 units), the losses of CHP amount to only 65 units, whereas the overall losses of conventional methods amount to 40 units for the heat production and 165 units for electricity generation – this yields an efficiency level of 80 percent versus 55 percent for conventional heat and power production.

CHP units achieve the highest efficiency gains when they are run under a fairly continuous heat (or cooling) demand and a constant heat-power ratio, for example in hospitals or on smaller manufacturing sites. Electricity production that exceeds the local needs can be fed into the central grid, while additional electricity demand can be met by the grid, too. For further heating or cooling,

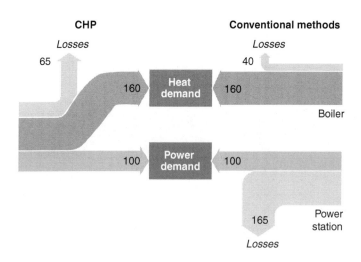

Figure 2.1 Efficiency gains of cogeneration, compared to conventional methods
Source: UK Department of Energy and Climate Change (2012).

boilers can be used. The International Energy Agency (IEA) (2008, 11) suggests an annual heating or cooling demand of at least 5,000 hours to benefit from the advantages of CHP-based energy provision.

The small-scale application of CHP in a decentralized energy supply hinges either upon a steady heat sink – like in hospitals or swimming pools – or upon an intelligent business model to financially exploit the efficiency despite daily or seasonal weather-related fluctuations. While the commercial viability of larger cogeneration units can often be taken for granted, our analysis focuses on the smallest segment of CHP units, the so-called micro CHP units, which are used in residential settings. Depending on the definition of "micro," their range typically does not exceed 10 kW_{el} and can be as small as 0.2 KW_{el} – sufficient though to replace a gas boiler in a residential central heating and hot water supply system, and to generate electricity that satisfies average household demand.

In this micro setting, internal combustion engines, in particular diesel motors, have been the predominant technology. However, a number of alternative technologies have been introduced recently, including fuel cells, turbines based on the Rankine cycle, and Stirling engines.

Fuel-cell technology creates chemical energy due to the oxidation of hydrogen. It produces heat during the chemical reaction, and is therefore an adequate technology for cogeneration. A number of companies currently develop proton exchange membrane (PEM) fuel cells, including Baxi Innotech, Vaillant, and BASF in Europe, as well as Toyota, Toshiba, and Panasonic – among others – in Japan. Most of the technologies are already in an advanced development stage with field trials. For example, the German company Vaillant has developed a PEM fuel cell with an electric output high enough to be used in multi-family homes and small business applications. The electric efficiency amounts to 35 percent and the overall efficiency reaches 85 percent.

Another alternative to the conventional combustion engine is the Stirling motor. Its fairly complicated technology is compensated for by the fact that any kind of sufficiently large heat difference can be transformed into mechanical energy, irrespective of the source of the heat difference. A variety of energy sources, including regular fossil fuels, biomass, solar and geothermal energy, waste heat, or even nuclear energy, can thus be used to operate it, making it a potentially promising technology. The use in CHP units has been explored over the past couple of years because of several advantages. In particular, the Stirling engine has a low noise level compared to internal combustion motors, because it runs smoothly and does not require explosions, like internal combustion motors, to operate. An increasing number of big companies as well as smaller start-ups promote Stirling engines for CHP use. For example, a company from New Zealand, WhisperGen, offers two Stirling products, the first one being an on-grid, heat-demand-led micro CHP system

Box 2.1 **Heat and power cogeneration at the World Expo 2010, Shanghai**

In 2008, China's share of cogeneration in total generation amounted to 13 percent – similar to that of Germany. The IEA estimates that this figure could rise up to 28 percent by 2030 under the right policy schemes implemented by the Chinese government. Most of this potential is linked to large-scale cogeneration. However, distributed generation based on decentralized combined heat and power may also be an environmentally sound option. In a study on the total technical potential for clean decentralized energy and combined heat and power, the World Alliance for Decentralized Energy calculated a 19 percent overall energy reduction and a 33 percent decrease of CO_2 emissions for four Chinese provinces and Shanghai until 2030.

During the World Expo 2010 in Shanghai, China, a heat and power cogeneration demonstration unit with a Stirling engine was set up. The 184-day demonstration and operation of the Stirling engine was part of the "best city practice area" for providing a good interpretation of the idea regarding energy-saving and low-carbon cities.

The demonstration of heat and power cogeneration of Stirling engines focused on the distributed power supply system, which uses a Stirling engine as a prime mover. Being fueled by natural gas, the heat and power cogeneration system provided 50 kW of electricity for the World Expo Park. It also used waste heat to produce 1.5 tons of hot water per hour for the dining center in the park.

Figure 2.2 The China Pavilion, an energy-saving building at the Expo 2010 in Shanghai

Source: Weinmann (2012).

The heat and power cogeneration system consisted of a 50 kW power generator set that uses a gas-based Stirling engine, a chassis-mounted body, a control and intelligent power gridding device, and a 12-cubic-meter hot-water tank. Since its successful application to the power supply of the World Expo, the distributed power supply system has contributed to low-carbon construction and new energy applications. It is envisaged that decentralized cogeneration will be more widely used in city construction to achieve a "better city, better life" in the future.

(*Sources*: WADE, 2010, IEA, 2009, Ding and Huihuang, 2012)

designed for domestic home use, and the other one an electricity-demand-led, off-grid heat and power system for marine or small off-grid remote dwellings and applications.

The third new technology for micro CHP units is based on the Rankine cycle. It resembles a small, conventional generation plant with a turbine. However, global sales of micro CHP units using the Rankine process remain limited.

Being a niche segment with lobby groups that are not as powerful as those in the renewable energy industry, CHP supporters and manufacturers in many countries have only recently succeeded in convincing political decision-makers that more pronounced incentives are environmentally and economically beneficial.

In the United States, a number of financial incentives are available in individual states and municipalities, including feed-in tariffs, rebates, and loans. On the federal level, the Energy Improvement and Extension Act of 2008 significantly expanded federal energy tax incentives and introduced the CHP investment tax credit. The American Recovery and Reinvestment Act, passed in early 2009, expands and revises tax incentives for CHP and provides funding opportunities for CHP and waste-energy recovery.

In the EU, member states pursue individual approaches to promote CHP. The UK government, for example, introduced a reduced rate of value-added tax on the purchase of the product and a generation tariff, whereas Ireland installed a feed-in tariff and France, a tax rebate. The Danish Partnership for Hydrogen and Fuel Cells and the Japanese "ENE FARM" project, under which Panasonic and Toshiba jointly market their products, have been established for the specific support of stationary PEM fuel cells in micro CHP units.

The German parliament ratified a new micro CHP act in 2008 with a target of doubling the CHP share in electricity production. The act renewed a bonus system, focusing on new installations being brought into operation before the end of 2014, including a bonus on electricity that is fed into the public grid or directly used in the industry. The program actually turned out to be so successful that it was halted because the cap had already been exceeded by 2009. However, the German government renewed the program for micro CHP units up to 20 kW_{el} with subsidies ranging from €1,500 to €3,500.

LichtBlick, a renewable energies retailer, launched a CHP program in 2010 based on micro CHP units developed by Volkswagen (VW). Ralph Kampwirth of LichtBlick comments on the regulatory support that micro CHP units have received in Germany:

> In formal terms, there is still the objective adopted by the last federal government in 2008 whereby 25 percent of power production was to come from CHP units by 2020. Initially, a budget block was placed on the

so-called mini CHP incentive program in early 2010 so that investments in CHP up to 50 kW$_{el}$ were promoted. Following Fukushima, the German government has set up a new initiative that supports micro CHP units with €3,450 per unit in the investment costs. In addition, the subsidies for electricity generation by micro CHP units have been increased from 5.11 to 5.41 cents per kWh for 10 years or a total of 30,000 operating hours. This is beneficial for our strategy, since our micro CHP units only run 1,500 to 3,000 hours per year. Even if we clearly did not rely on public funding in our cost calculations, we are satisfied that political decision-makers fund solutions that favor intelligence and flexibility in the system. (Kampwirth, LichtBlick)

The future generation portfolio, based predominantly on fluctuating renewable energies, will increase the necessity to re-initiate some type of state aid for flexible generation units like CHP:

> If wind and sun are in the future to provide the bulk of energy, we need solutions for making more generation capacity available. The German electricity market is not yet well positioned on these points. Expansion of the grid is therefore indispensable. But how much expansion is needed depends on how much flexible generation like CHP units is possible. Funding for flexible generation ultimately saves money. (Kampwirth, LichtBlick)

For the final customer, the attractiveness of the technology crucially hinges on the cost competitiveness, which also depends on the general level of electricity tariffs. According to LichtBlick, an omission of the tax components of the tariff could be used to enhance the commercial viability of CHP units. A reduction in regulatory uncertainty is crucial for the further development of cogeneration:

> CHP units become viable when electricity prices increase. Taxes and levies are the most expensive components of the electricity price. Exemptions for CHP plants may be an option to promote the technology. What is important today is a clear timetable and appropriate framework conditions. (Kampwirth, LichtBlick)

Within the current regulatory context, natural gas should be promoted more strongly, according to LichtBlick. It can serve as a transition technology that provides low-carbon primary energy until a 100-percent-renewable supply is achieved. Kampwirth considers a subsequent, gradual transition to biogas as being inevitable, though.

A sensible energy concept must ensure the expansion of renewable energies as well as clarify the issue of grid, storage, and flexible energy generation. In our view, natural gas or biogas could form the basis. Natural gas is the fossil fuel considered the bridge into a new era and a resource to complement renewables for as long as necessary. Therefore, its market development should be promoted. LichtBlick has set itself the goal of generating 100 percent with renewable energies by 2050. By then, the age of natural gas for electricity generation will have passed. As an alternative, we are betting on biogas. (Kampwirth, LichtBlick)

Radu Anghel, former CEO of Greenvironment, a company that operates both micro turbines and CHP units, emphasizes that subsidies for CHP units in Germany are limited:

Even today, CHP subsidies only cover 30,000 hours. Beyond that, operators have to be inventive how they can establish a viable business model. For example, they can pool the plants via a contractor. (Anghel, Greenvironment)

Nobuo Tanaka, the executive director of the IEA, considers the major obstacles for a successful expansion of CHP units to be that CHP benefits are not (yet) recognized in most national greenhouse gas regulations, and that there is a lack of information about cost-savings and other benefits. In addition, Tanaka remarks that developing countries are only slowly beginning to see the potential benefits of CHP (Tanaka, 2008).

Andreas Prohl, Member of the Board of Directors of gas utility GASAG, comments on the bureaucratic hurdles and red tape still hindering self-producers to get connected:

With regard to the general policy environment, action definitely needs to be taken and much needs to be simplified. At present, if one wishes to connect a small system up to the grid and receive a feed-in remuneration or to get the gas tax refunded, there are 17 application forms to be filled in. That is asking too much of most customers. (Prohl, GASAG)

More than 80 percent of total global electric CHP capacity is located on industrial sites in the food processing, pulp and paper, chemicals, metals, and oil-refining sectors, where process-related heat or cooling requirements guarantee a continuous demand for the thermal output (IEA, 2008, 11). While industry has been applying the technology for a long time and many municipalities are relying on large district heating networks fueled by cogeneration plants, only recently has there been a wave of new and relatively cheap CHP technologies – often initiated by local communities or environmentally minded neighborhood organizations – that have broadened the market for small units run by non-industrial customers, such as commercial and institutional users. COGEN Europe, the lobbying association of the combined heat and power producers, assumes that by 2050 more than a quarter of all electricity could be generated by combined heat and power (COGEN Europe, 2011).

Even though the technology for CHP at a household-size level is available and mass production is about to be launched by companies such as Volkswagen and Toshiba, the economic viability is often questionable.

Between 2006 and 2010, global sales of micro CHP units fluctuated between 20,000 and 25,000 units per year. In 2009 and 2010, a substantial fraction of sales of micro CHP units based on internal combustion technologies was substituted by PEM fuel cells, mainly due to a large-scale investment program in Japan. Despite the fact that numerous countries have set up policy measures to promote applied R&D and market rollout of micro CHP units, without Japan's massive policy push toward fuel cells (see Box 2.2), sales would have even declined.

Figure 2.4 provides an overview of annual micro CHP sales by technology. Players in the market have largely differing views on the prospects of micro CHP units, though. LichtBlick, for example, is pursuing a target of 100,000 installations within the next years, and even sees a large potential beyond that figure:

> LichtBlick received 40,000 applications for its CHP program in the first few months. In addition to residential households, we install our CHP units already in all types of buildings such as kindergartens, blocks of apartment houses or, where possible, small commercial premises. The demand in coming years for renovation and new installations will be so high that it would not be possible to satisfy the market. This is the reason for the optimistic goal of 100,000 CHP units.
>
> At this point, we see ourselves as an essential player because we could reach a generation capacity of 2,000 MW with 100,000 home generators. That is equivalent to the capacity of two nuclear power stations with all the mentioned advantages that we have in terms of safety, flexibility, etc. (Kampwirth, LichtBlick)

The market growth crucially hinges on the right business model:

> Hitherto market development has been very restrained and characterized by a small-scale strategy. Projects such as ours may have injected dynamism into the market. It would be desirable to have more innovative and affordable solutions that are attractive to the consumer. (Kampwirth, LichtBlick)

Box 2.2 "ENE FARM" – commercialization of a residential micro CHP unit based on fuel cells

Micro CHP units based on fuel cells use have been promoted by the Japanese government with its ENE FARM project since 2009. In cooperation with electronic equipment manufacturer Panasonic, the utility Tokyo Gas has sold over 10,000 ENE FARM units in the three years between the launch of the scheme and April 2012. The fuel cell has an electric energy efficiency of around 40 percent. The efficiency of the heat recovery system amounts to 50 percent. An average household can save about 1.5 tons of carbon dioxide per year with the device.

The units are based on low-temperature fuel cells called PEM fuel cells that operate at temperatures of 60 to 80 degrees Celsius; they have clean exhaust gases, and low noise and vibration levels. For residential applications, Panasonic further simplified the design of its PEM fuel cells and downsized the stacks that generate electricity. More specifically, it changed the shape of the fuel-cell to a vertical design and connected it with the hot-water storage unit, such that the system would fit on 2 square meters. With a 30 percent reduction in the number of required parts and a 20 percent reduction in weight, Tokyo Gas and Panasonic could reduce the retail price by approximately 20 percent. Still, with a price tag of more than €27,000, which includes tax but excludes the installation fee, the technology hovers at the upper end of an acceptable price range for a residential heat and power unit. However, Tokyo Gas claims that residential consumers can achieve annual savings up to €600 on their electricity and heat bill and aims to have sold more than 16,000 units by the end of 2012.

Panasonic plans to expand into the German market with its product in 2015.

Figure 2.3 Residential fuel cell unit used in the ENE-Farm initiative
Source: Panasonic (2012).

(*Sources*: Tokyo Gas, 2011, Tokyo Gas and Panasonic, 2011, Tokyo Gas, 2012)

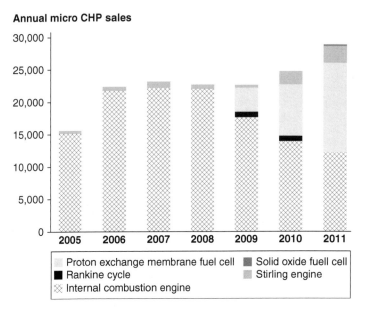

Figure 2.4 Global micro CHP unit sales by technology
Source: Delta-ee (2012).

Compared to global sales figures of around 23,000 units in 2010, this estimate would lift the whole industry to an entirely different scale. Until spring 2012 LichtBlick had installed 420 units (Schlandt, 2012b).

Representatives from other companies remain more skeptical about the market. For example, Manfred Greis from heating installations manufacturer Viessmann – a company specialized in small heating systems and active in the micro CHP market since it acquired micro CHP specialist ESS in 2008 – sees a moderate but growing market potential.

> The market volume in Germany is perhaps around 500 units a year. The figure can and will incrementally increase. (Greis, Viessmann)

Jürgen Willand from car manufacturer VW is equally skeptical about the commercial viability of micro CHP units. He expects valuable insights in business models from the current cooperation with LichtBlick.[1] A significant market potential can be expected, but only above a minimum threshold that limits the economic viability of micro CHP units.

> In the years ahead, the market for residential heating installations that need to be replaced will exceed the currently planned production volume by far. We know how to proceed on the technical side, but the business

model still needs to mature. The smaller the CHP unit becomes, the less electricity it will produce. The question is whether the LichtBlick model for small CHP units is really attractive to potential customers and makes economic sense. In this regard, the experience that we are building up with today's concept and with the current CHP units will provide important insights. In any event, there are heat sinks that are markedly smaller than what is economically feasible for our CHP unit. A CHP unit that has half as much output as ours will probably only be half as expensive and correspondingly capable of being operated less profitably because the electricity output that can be sold is too small. A CHP unit would then perhaps have to run for ten hours to be able to deliver the quantity of energy needed for one hour of feed-in. This does not yet add up to a viable business model. (Willand, VW)

The potential market success of a CHP unit – encompassing island systems as well as centrally coordinated CHP units, so-called virtual power plants – depends on the socio-economic and geographic context:

If you can interconnect a large number of generators that are each operating efficiently, you will obviously have a high overall level of efficiency. The issue of CHP is in large part an issue of networks and communication. Using CHP units, you can substitute for large generators with poor efficiency and provide the necessary reserve output. However, it depends on the infrastructure in a particular region. A CHP unit can operate very well in remote areas, even without a network. The more industrialized a region is, the better the infrastructure will be and the better the possibility for interconnecting CHP units and steering them centrally. (Willand, VW)

Eckhardt Rümmler, Senior Vice President Strategy & Corporate Development of German energy utility E.ON, does not yet see the market potential of micro CHP units, but expects it to develop once mass production kicks in:

To date many distributed energy technologies cannot live without initial subsidies or even permanent subsidization. But how successful can a technology be if the subsidies are withdrawn? Is it possible to create a business model on this basis?

After the year 2000, we thought about whether we should grasp business opportunities solely relying on subsidies. With hindsight, it was a mistake not to enter the wind business earlier. We had to learn and accept that part of the income stream in the energy sector can come from subsidies. Certain technologies are even only realized when subsidies for them are available in order to make them economically viable. However, it is always the target to increase the efficiency of a technology in order to go without subsidies.

When we established our Climate and Renewables company subsidiary in 2007, we considered whether distributed energy would also deserve to be shifted to a separate business unit. But we decided against such a move at that point of time. Apart from photovoltaics, which is still being hugely promoted by subsidies and thus questionable in many European countries regarding their macroeconomic benefits and costs, we did not see sufficient new business opportunities in distributed energy justifying already a separate business unit in 2007.

Since then, the market for distributed energy has emerged further and we have used the time to deepen our market understanding and to assess business opportunities. For example, we take a look at micro CHP units, which can replace old gas-fired boilers. The current CHP units are more like collector's items for technology-savvy customers. Yet the learning curve is pointing sharply upwards. As soon as large-scale manufacturing sets off, the models will also become economically viable. Also, many renewable technologies show steep learning curves. For example, prices for photovoltaic modules have decreased between 2008 and 2011 by more than 50 percent. (Rümmler, E.ON)

The long-term potential of micro CHP in a carbon-constrained world may be limited, though, as Alexander Voigt, founder of island solutions startup Younicos, remarks:

> For the next 20 to 30 years, micro CHP units will make sense because they use gas far more efficiently than conventional heating systems. However, in the long term this technology will not bring us much closer to our emission reduction goals. Yet, taken in conjunction with decentralized storage and renewable energy generation like photovoltaics, it is an extremely exciting technology. (Voigt, Younicos)

Carsten Liedtke, speaker of the Board of Directors of municipal utility Stadtwerke Krefeld, expresses an optimistic opinion on the growth in the segment of small cogeneration. He expects further demand from commercial customers as well as small and medium sized enterprises:

> Smaller installations operated with gas or renewable energies will experience a renaissance. We can make use of this at the local level because we already have the technical expertise and, in contrast to large suppliers, we have a detailed knowledge of local supply and demand structures, and the

topography. We know our customers and can, for example, set up installations run independently by them. These concepts were already interesting for industrial customers with a demand of several tens of GWh a year. We operate such installations in Krefeld for large businesses in the food industry. But smaller firms with demand of little more than one GWh increasingly consider acquiring such installations, too. (Liedtke, SWK)

Larger utilities like the Berlin gas supplier GASAG see considerable technical potential in local electricity and heat production. According to Andreas Prohl, the contribution to overall supply will remain fairly limited, though.

We would like to mobilize and channel CHP generation. We think that has potential and expect around 4,000 MW in a period of 10 years if we include small installations and local district heating supplies of between 1 kW and 2 MW. Hitherto in Germany, we have installations with an output range of around 3,500 MW. And 4,000 MW is about 5 percent of the annual peak load, and does not therefore represent the ultimate solution for energy and electricity supply. Nevertheless, it is an important contribution for our sector and also manageable in terms of the investment risk: we only buy a production installation when the corresponding customer base is there. In addition, we can rely on decades of experience on the market. (Prohl, GASAG)

Manfred Greis from Viessmann expects that new players, including the big energy incumbents, will enter the market:

It is obviously possible that new players will enter a market that was relatively closed in the past. We had manufacturers such as ourselves that supplied the market. Energy suppliers with completely different ideas enter the market. In addition, there are completely new players, at least in this area, such as VW, that have the motor technology and therefore seek access to the CHP segment.

The large energy suppliers, not only LichtBlick but also E.ON, RWE, and at least two others, are also obviously deciding their strategies for the future. They know one thing for sure: they will sell less electricity and gas. Because if society wants efficiency – as we must – then the volume of energy will decrease. But they nevertheless want to maintain their turnover and, above all, their profits. That means they must either increase their prices so that they earn the same return on lower energy sales or they must create new business models. For that reason, new structures will undoubtedly be put in place. (Greis, Viessmann)

Johannes Hengstenberg, founder of co2online, a not-for-profit consulting company offering online advice in energy and building efficiency, comments that increased competition may help market development.

> The marketing director of a cogeneration manufacturer told us that a market for micro CHP units has only started to develop since several suppliers came new to the market. Currently all parties benefit from the emerging competition. (Hengstenberg, co2online)

Exploiting scarcity pricing in the wholesale market: virtual power plants (LichtBlick)

LichtBlick was founded in 1998, right after the liberalization of the German electricity market, with the objective of providing "clean energy for a clean price." By 2003, the company had attracted more than 100,000 primarily residential customers; by mid-2011, around 600,000 customers were registered. With this customer base, LichtBlick is the largest German electricity retailer that is independent of any of the large utilities.

> LichtBlick has entered the market as an energy trader, hence, the company purchases eco-electricity and sells it to customers. But from the beginning, the founders had the intention of entering progressively into generation. With 600,000 customers, we have reached the critical size at which we can enter energy generation, and we also have the financial resources to do so.
>
> The market for micro CHP plants requires intelligent strategies, given that the full economic and environmental potential of CHP can only be realized when both heat sinks and electricity demand are available. (Kampwirth, LichtBlick)

The company had the idea to use price fluctuations in the German electricity wholesale market to generate sufficient revenues for micro CHP units to be profitable even in individual households.

> With the expansion of renewable energies, the energy system faces completely new challenges. Central issues of Germany's new energy policy such as grid expansion and electricity storage are already being discussed. The construction of intelligent back-up power stations has for a long time been neglected, but now the federal government has realized that support for additional generation capacity may be needed. We recognized early on that we would run into an energy supply problem due to the rapid growth of renewable energies. Renewable energies such as wind and sun are generated when the weather conditions are right and not necessarily when the consumer needs electricity. We need storage possibilities when there is too much sun or wind or a flexible back-up system when there is too little.

The current fleet of power stations is structured as follows: we have the base load, which is met by nuclear and coal-fired power stations, and then a peak load. If we look at forecasts, this system will be outdated soon. With the expansion of renewable energies, we will have enough power to meet 100 percent of our electricity demand via renewables as early as 2030. Simulations show that we would then have to react flexibly and switch off all the non-renewable power stations. However, in the case of nuclear power stations, only 50 to 60 percent of the output can be taken from the grid because of technical restrictions. After a complete shutdown, it takes at least two days to reconnect with the grid. Furthermore, there is a safety issue. Accordingly, since the necessary balancing energy may not always be available in the market, we have to find other solutions.

Our micro CHP units, which we refer to as home generating units, are based on the idea of offering an extremely flexible electricity supply. Start-up takes only 60 seconds. With cogeneration, we use an established technology. Even though the energy source used – natural gas – is a fossil fuel, it is used very efficiently. Our efficiency rate is more than 90 percent. By way of comparison, an old coal-fired power station manages only 30 percent, a new one perhaps 45 percent. Natural gas only releases about half as much CO_2 as coal. In terms of climate and efficiency, we are already very good. The electricity from our generators is recognized as eco-electricity by environmental associations and the usual green electricity certification bodies. We believe that gas in CHP constitutes a bridge technology because it is comparatively climate-friendly and extremely flexible.

CHP units are difficult-to-market niche products for amateurs and entail high investment costs. For that reason, there are only a few CHP units in Germany and very small suppliers. We therefore had to develop a model that can function economically and is suitable for the mass market.

A normal CHP unit is either electricity-led or heat-led. LichtBlick developed a completely new concept: our micro CHP units generate electricity at peak loads, when there is strong demand and high prices. However, we can still comply with the customer's heat requirements – even if heat is needed during other times than the peak-price phase when we let the cogeneration unit run. This works because we connect the CHP to residential heat storage units. At times when LichtBlick is running the installation, the heat is temporarily stored in the customer's basement. In the case of a two-family house, these heat storage units can hold 2,000 liters and more. In principle, the customer draws heat from this storage unit and not directly from the generator. In this respect, we harness the fact that heat can be stored easily, unlike electricity. We use the heat storage unit as a buffer, so that we can generate electricity in a targeted way.

> LichtBlick rents the customer's basement and installs a micro CHP, which we also control and operate. The customer obtains the heat from the generator's storage unit. This is actually the only service he pays for, while we take the electricity and feed them into the local grid. The generators are activated by the Hamburg control station using mobile technology. The generators are designed in such a way that they operate for between one and seven hours a day. This is a further difference as compared with classic CHP units, which usually operate for around 6,000 hours a year. In this way, we provide the electricity market with large volumes of energy in a short period and which the market needs to compensate for lack of feed-in of renewable energies. Because we use techniques of crowdsourcing and coordinate these small generators like a swarm of fish, we call this concept "swarm electricity." (Kampwirth, LichtBlick)

Kampwirth characterizes the target customers for this business model and explains the financial conditions:

> Customers who are looking for a new gas heating system are our main target group. Here we are competitive and as a rule markedly less costly than conventional heating devices. The connection fee for a micro CHP starts at €5,000; installing a conventional heating device costs at least €8,000. At least, a further 20 percent can be saved on operating costs because the customer pays only for the heat and also receives a bonus for the feed-in electricity. LichtBlick buys the gas and also takes charge of service and repairs. The 40,000 customer inquiries received over the last six months and the interest this demonstrates in our product confirms our optimism that we can also reach our target of 100,000 installations. (Kampwirth, LichtBlick)

After having concentrated on Hamburg in the initial phase of the project, LichtBlick has plans to expand in other German urban areas:

> In autumn 2010, we installed the first customer units in Hamburg and are now in the rollout phase. At the moment we install around 10 new units per week. The focus is on Hamburg for the time being, so that we can gain experience in the market. The product is completely new, with innovations such as intelligent heating control. This also means that the product quality is not comparable with traditional CHP units.
> With the information collected, we can then guarantee high quality as well as a rapid rollout in the mass market. Across Germany, we are already in contractual negotiations in seven regions. We are currently concentrating on agglomerations, but in the longer term, we want to progressively position ourselves more widely. (Kampwirth, LichtBlick)

The implementation of the LichtBlick business model has faced a number of hurdles, in particular skepticism of the grid operators:

> It was difficult to configure the technical pre-conditions of the product in such a way that it fits well into the market. For instance, we had to find a way how to measure the energy. Given that this market has been liberalized, LichtBlick itself is now allowed to operate intelligent control devices for the installations. For the local grid operators, this is a completely new concept – and not all of the 800 grid operators in Germany are enthusiastic. Some have even requested external electricity meters. However, we have managed to prevail. We regard ourselves as a good partner, but will not put up with everything and see ourselves to some extent as an engine of competition and liberalization in the market. We would like to actually implement what is legally valid and are in close contact with the Federal Network Agency in this regard.
>
> But there are also many positive reactions from network operators. As soon as we have installed a certain number of micro CHP units across Germany, our installations can constitute a complement to weather-dependent electricity generation from, say, photovoltaic installations, since we can make reserve or flexible energy available to distribution networks. If intelligent electricity is available, network operators can save costs for the grid expansion that would be necessary as a result of volatile input from renewable sources.
>
> In media terms, we have managed to bring CHP units out of their niche by communicating through careful choice of formulations. Interestingly, the concept has still been misunderstood by a number of experts. There are also some critical views, for example, from the heating manufacturers who are observing very closely what we do. But we work against such reservations. (Kampwirth, LichtBlick)

Two of LichtBlick's core competences are the company's expertise in information technology and in legal matters.

> LichtBlick has always made certain things itself, such as the IT, which is extremely complex in the electricity and energy market. For LichtBlick, this is an intensive practice with high internal know-how. Our IT employees are process managers who have a broad practical knowledge and develop appropriate solutions. We work almost exclusively with systems originating in-house. The entire control system for the home generator and the underlying software were developed by our specialists. Energy-trading and portfolio-building in the electricity and gas market are also managed by us. (Kampwirth, LichtBlick)

In a competitive environment, legal matters become increasingly important, especially in the interaction with the local grid operator and the federal network agency.

> We have a well-functioning legal department, which knows how we can secure our rights. We consider this an important asset, since we see ourselves as a core driver for competition and liberalization in the market, with the desire to implement what is legally achievable. We have a great deal of experience. All partners in the market know that LichtBlick is a reliable partner and not one to mess around. We are always in close contact with the Federal Network Agency because we have an innovative, challenging product and want to secure our revenues. Many network operators react very positively, and we have already had promising experiences. (Kampwirth, LichtBlick)

For the installation of the micro CHP units, LichtBlick trains and cooperates with local technicians.

> LichtBlick does the rollout and takes care of all sales activities. We occasionally work with sales partners, too. However, since we cannot coordinate installation of the home generating units on our own, we seek out regional firms as partners. We started with a small technical firm in Hamburg with which we developed a training concept. Each new artisan firm receives training. All processes that we develop can be scaled up and are in line with the goal of mass production. To this end, we build upon our experiences in the eco-electricity and gas market.
>
> By May 2012, we hired 150 new employees and now operate around 500 micro CHP units. We still stick to our target of installing 10,000 units per year, but we have to ensure the quality of the installations and have to train our partners, which of course takes some time.
>
> Many firms are interested in a cooperation with us and have approached us to offer a service partnership. We have built up so-called service partner management. This means that we have service partner managers who accompany our service partners and the local technicians. They seek out and train craftsmen, mediate in problem cases and are the first point of contact. They are then responsible for installation, maintenance, and support services for the micro CHP units, which means that we can offer our customers a rapid, high-quality response.
>
> We monitor our partner firms for installation and maintenance. Every house is different so that modifications are necessary during installation. (Kampwirth, LichtBlick)

According to Kampwirth, a change in the attitude of German energy incumbents toward a decentralized energy supply can be observed:

> Until recently, the large energy suppliers simply had no interest in decentralized generation. However, in the last year the situation has fundamentally changed. It would be wrong to claim that LichtBlick will determine the situation alone. But we are clearly happy to have been first in the field. (Kampwirth, LichtBlick)

Even if more competitors enter the market, Kampwirth is confident that his company remains in a comfortable position due to its innovative strategy and portfolio of future business areas.

> At the moment we have a certain lead because we were the first on the market. Since we communicated the concept and the partnership with VW, more energy suppliers are looking at the area of CHP mini-generators, including Vattenfall and RWE. There is cooperation between Honda and Vaillant where the concept is not uninteresting but clearly different from what we are doing. Even Deutsche Telekom has announced to enter the market for micro CHP units. However, none of our competitors has as far-reaching objectives as we have. (Kampwirth, LichtBlick)

In the cooperation between Honda and Vaillant that the interviewee mentions, Honda provides the motor, while Vaillant is responsible for the hydraulic parts, the system control, and the links with the household appliances (Honda, 2009).

> In terms of communication and marketing, we are well positioned and have a market lead, but we are not so naïve as to believe that this will always be the case. We are not afraid of competition. We want to broaden the market segments and develop various further pilot projects with our partners. This ranges from heat supply for larger residential settlements to publicly owned properties, etc. We are optimistic that we will have good rollout in future and already have a wide spectrum of projects at all stages of development. (Kampwirth, LichtBlick)

German CHP manufacturer Viessmann propagates a more traditional business model than LichtBlick. Its profitability primarily relies on a constant demand for heat:

> The idea of LichtBlick and Volkswagen is interesting. They use a great terminology with the swarm intelligence. In my opinion, the LichtBlick model is

a public relations issue, though: "Here comes the home generator." I do not see it that way. A four-cylinder, two-liter motor is oversized for a regular one- or two-family house. Its place is in larger buildings, for instance in stately homes with heated swimming pools, small hotels, etc. CHP units of this size have been on the market for a while – LichtBlick and VW are not the first.

I cannot imagine that CHP will really contribute on a larger scale to our electricity requirements in the future because that would mean that the heat would not be used. We have to look at the process led by heat sinks and aim for the longest possible running times. A CHP should only be configured to the size that corresponds to the heat requirements in the local context. The CHP should operate for a total of at least 5,000 or 6,000 hours per year. It would obviously be ideal if a CHP could run for all 8,760 hours in a year but we will not achieve that. We configure it somewhat smaller but will then always build an additional peak load heat generator.

We naturally have to move quite some way from the purely central structures of power generation that we have known to date. They were built with a completely different mindset. Often, they were located near the banks of a river so that they could be supplied with coal on the waterway and above all to get rid of the heat. Locations were chosen such that cooling was possible. Today the situation has turned round and we use this heat. But then we need to have a demand for it, otherwise it makes no sense. That is ultimately the restricting factor for CHP. (Greis, Viessmann)

Representatives from large companies, both energy incumbents and diversifiers, remain skeptical. Eckhardt Rümmler of E.ON considers it an interesting niche product, but doubts its long-term financial viability:

The LichtBlick model is in reality only combined heat and power cogeneration in a smaller plant, exploiting differing prices in the wholesale market. But how much longer will there be this kind of wholesale market or to whom can sales be made? Nevertheless, it is an intelligent niche product. (Rümmler, E.ON)

Micro CHP units may even prevent investment into more efficient technologies and create a suboptimal lock-in, according to Michael Weinhold, Chief Technology Officer of the Energy Sector with multinational manufacturer Siemens:

Those firms that are trying to create a business case on the basis of decentralized combined heat and power plants still encounter difficulties. Some of these small power stations generate lots of heat. Someone who has such a facility in his basement is not interested in further insulating his house. In the worst case, the infrastructure in our basements is not only inefficient by itself, but is an obstacle to boosting overall efficiency. It is far more efficient and also effective if we rely on highly efficient large-scale power stations where we also use the heat. In this way the energy source is exploited better by a factor of two to three than it is by installing a small basement power system that uses electricity inefficiently. (Weinhold, Siemens)

In energy-efficient houses, heat pumps are a serious competitor to micro CHP units because heating requirements have become negligible due to better insulation techniques.

Instead of burning a fossil energy source merely to produce heat, we should produce electricity highly efficiently and then increase this electrical energy by a factor of three to four, especially with highly efficient heat pumps.

The emphasis on electricity and electrical efficiency is important because in future above all electrical energy will be used in buildings as a result of better insulation and low temperature and zero-energy houses. It occurs quite often in new housing projects that no gas pipelines are being installed because the buildings are so efficient that they do not need any gas for heating but only a power line through which, with the help of a heat pump, they can generate the few thousand kilowatt hours that are needed. The typical single-family home built in the 1960s has an annual thermal heating requirement of 30,000 to 40,000 kWh. In contrast, a new building today needs well below 10,000 kWh. A low or zero-energy house only needs intelligent ventilation.

The German government still plans that cogeneration contributes by one quarter to the production of electrical energy in 2020, although electricity becomes increasingly important as a result of better insulation of buildings. Does it make sense to provide incentives for CHP units and even micro CHP units, even if this is an obstacle on the path to an energy-optimized infrastructure? (Weinhold, Siemens)

A couple of years ago, E.ON did not see the potential of decentralized energy generation. However, that perspective has changed recently, according to Eckhardt Rümmler:

We investigated the issue of distributed generation already in 2006 and assessed potential risks for large-scale conventional power generation. We clustered the issue into micro plants, mini plants, and larger plants, especially CHP plants. We analyzed how many Terawatt hours our large-scale generation plants may lose because of the operation of small and medium-sized plants and asserted that the amount was small and will remain small in the foreseeable future, even though the number of distributed plants was expected to grow continuously.

We will see a further emergence of customer-oriented distributed energy solutions in addition to the existing well-known centralized energy world with its large scale generation. Therefore, "distributed" and also the issue of "smart" have their clear part to play in our "cleaner and better" corporate strategy with distributed energy being one of our key development priorities. This affects both our established sales and heat business activities, as well as new business areas for E.ON, such as renewable distributed generation or even business model innovations such as virtual power plants. In order to deliver on our growth ambition in distributed energy, not only our Regional Units drive forward the implementation, but we established in addition in 2012 a new unit responsible for a selection of distributed energy initiatives to be commercially rolled-out across our markets.

In the case of new areas of business, there is always the possibility of a conflict arising with the main existing business as priority setting is challenging. Nonetheless, a company has to decide whether to launch activities within existing organizations or to ringfence them into a new unit. Thereafter, it has to reverse the organizational structure and to re-integrate the business unit into the organization at an appropriate time.

There are certain elements in the distributed area that enhance efficiency, and others that do not. Subsidizing technologies without a proper assessment of its long-term economic viability will not reinforce Germany's pioneering role. Other countries, such as the United States and the United Kingdom, are more pragmatic and unemotional. (Rümmler, E.ON)

Despite the reservations that Siemens and E.ON representatives express, two companies have announced that they will enter the micro CHP market much as LichtBlick has done. Swedish energy utility Vattenfall announced it would extend its micro CHP fleet, which was hovering around 50 in summer of 2012, to total an electric capacity of 200 MW by the end of 2013. The business idea imitates part of the LichtBlick model by selling electricity on the wholesale market during peak hours and buffering the heat for residential usage. By contrast, Vattenfall aims to also integrate heat pumps into the system, which use electricity to generate surplus heat when abundant wind energy is available. The units are centrally controlled from Berlin. The project team is composed of CHP manufacturer SenerTec, heat pump provider Stiebel Eltron, and Vattenfall (Stiebel Eltron, 2010, Schlandt, 2012a).

Moreover, German telecommunications incumbent Deutsche Telekom plans to install CHP units in residential settings, albeit with up to 1.2 MW capacity for larger settlements. However, the company will sell its units to municipal utilities after installation and will not be actively involved in operation and maintenance. It is partnering with CHP producer Motoren AT and IT company GreenCom Networks (Schlandt, 2012b).

One possible obstacle to the LichtBlick model is the drying-up of the electricity spot markets. Due to a higher share of feed-in from must-run plants like photovoltaic panels and wind rotors, liberalized electricity markets are being increasingly re-regulated. Volumes on spot markets decrease, trading becomes illiquid. If price-spreads on wholesale markets became erratic and uncorrelated to actual scarcity, the profitability of swarm intelligence could face a severe challenge.

Manufacturing synergies and economies of scale (Volkswagen)

The cooperation between LichtBlick and VW brings two players together that do not have long-term experience in energy generation. However, the complementary fields of expertise provide a sound basis for cooperation.

> In early 2008, we put together a team of engineers for the project. When we had developed the swarm electricity concept and were considering what the home generating unit should look like, we wondered who an adequate partner could be. To enter the mass market, we needed serial production. All past manufacturers of CHP units made only small numbers by hand. We heard that VW had already developed a CHP on the initiative of the workers' council. VW's purpose was to identify a different use for motors in order to diversify. For instance, VW also builds engines for fork-lift trucks and sports motor boats, but only in marginal quantities. So we got together with VW, presented our concept, and explained what we would need for implementation. We started a partnership to develop the micro CHP on the basis of the VW generating unit "EcoBlue." With this partnership, VW has placed great trust in us. Both VW as a global group and also LichtBlick as a smaller but more flexible and innovative energy group are collecting experience and learning a lot from each other with the project. VW is a very competent installation developer and builder with very high quality demands and know-how: our home generator has a noise level of less than 50 decibels thanks to very efficient insulation and as a result is not much louder than a refrigerator, which increases the attractiveness for the customer. (Kampwirth, LichtBlick)

The experience of VW in mass production facilitates cost degression of the modules:

> We have found an excellent partner in VW. Since VW produces our generators in series and not, as is usually the case, in manual production, prices

have noticeably come down, and the attractiveness of the product has increased. (Kampwirth, LichtBlick)

Since the 1980s, Volkswagen has gained experience with motors used in cogeneration.

> It is estimated that the idea of cogeneration is 60 years old. Volkswagen is, of course, not the first firm to have explored the field. Many others have also done so, including other automakers, but have rejected the idea as not being feasible. Already in the 1980s, Volkswagen was on the way to placing a CHP unit on the market. Declining energy prices in that period then meant that the idea was no longer attractive, and the project was put on ice. Nevertheless, thought was already given years ago to develop a CHP plant, but it was not yet economically feasible. The costs would simply have been immensely high. Our competitors currently operating on the market show this, too. Even if the technology is not highly sophisticated, units often cost €20,000 and more. The average house builder cannot afford that and will look for an alternative.
>
> Despite this, we have invested a great deal of energy in development over a long period of time, in order to drive the concept forward – not as a core project, but still investing time and money. We have been running the prototype of a CHP unit here on-site since 2008. To date it has been providing heat for the washing facilities with a relatively large demand for hot water.
>
> Because we wanted to show that we have this expertise, we have been represented with the CHP unit at various trade fairs. This is how we came into contact with LichtBlick. LichtBlick presented a business model to us and we confirmed that we can implement it. It became a full-fledged project in September 2009, on the day that we announced our cooperation with LichtBlick. (Willand, VW)

The company was able to benefit from technological synergies with its car manufacturing expertise:

> Although this is a completely new product for Volkswagen with correspondingly high requirements, it did not have to be completely invented and developed from scratch.
>
> A CHP unit actually has many parallels with an automobile: we have a generator, which corresponds to the dynamo; the gas extractor is the exhaust; the heat exchanger is the radiator. I also have a control unit in the automobile. The motor also originates from the automobile. If you look inside a CHP unit, you will also recognize many things that you also know from, say, the Golf such as an oil sump, oil equalizing reservoir, and such like. Volkswagen has all these competences in-house. One could claim that we can serve ourselves from the corporate toolbox. VW makes heat

exchangers in Hannover and exhaust systems in Kassel. The steel plate for the cover is produced in Sarajevo. But we can also benefit directly owing to group procurement from suppliers, for instance for the generator and ball bearings, which we purchase from other manufacturers. (Willand, VW)

Compared to micro CHP units from other manufacturers, the system developed by VW is characterized by a high degree of integration of all components within a single module.

> Our CHP unit is modular in construction, modular in design, and modular in implementation. The basic module is the motor with generator, gas extractor, and heat exchanger, the hot core of the CHP system. We then have a hydraulics module, in which all water pumps, pipes, etc., that are normally installed on the wall in a domestic installation are integrated in a compact manner. This is all very atypical for the sector, but precisely what creates the attraction of our CHP unit. The third module is the electrical module containing all control functions. With us, there is no separate switch box as is normally the case in the sector. We come from the automotive industry and are used to having to build many control functions into a small volume. We therefore know how to miniaturize. Before systems are installed, the generator is removed from the basic module and the installation is broken down into four sections so that it can be placed in the basement. It can then be moved more easily. (Willand, VW)

During the development phase, involvement and feedback of future customers was essential.

> We agreed right from the start that we would involve our customers in pilots at a very early stage. We accelerated our typical development processes to reach the goal more rapidly.
> The most important element of a CHP unit is the motor, a motor that is supposed to work for 30,000 hours. Yet such an endurance of 30,000 hours is not usual in the passenger car sector and was new territory. First, we had to ensure that our motor was actually capable of running for so long. Time cannot be compressed, that is the problem. We can operate for only around 9,000 hours a year. If a product has to work for 30,000 hours, you either need to be naïve – or to deploy a professionalism that enables you to gain a picture of how to make a product durable despite shorter test periods. That was our task. Hence, we first let the motors run with only parts of the periphery system, which allowed us to delay tests of the complete product until the next prototype generation.
> We reached an arrangement with our customers and LichtBlick to deliver so-called field test CHP units. These were handmade prototypes that were

supposed to demonstrate all the functionalities. They were installed in real houses with real people with an interest in such a system. In this way, we had access to our partner LichtBlick's end customers and already had to deliver a high-quality product because the customers would have otherwise faced the risk of having to take a cold shower. Incidentally, that did happen once or twice. That is when things heat up. It was also a new situation to us where we had to deal directly with the end customer – a challenge and a completely new field, but an absolute necessity to meet the quality requirements in a deep and wide pilot operation.

We also learned that satisfying acoustic attributes of the units here on site were not adequate in a real basement. In our acoustics facilities, we have dampers but in the basement there are just bare walls. Not only is the sound different – that might perhaps be acceptable – but, due to resonances, the system became louder than it should be. We then worked hard and have now reached the point where we are top of the class in comparison with everything that is known about CHP units, both in the acoustics hall and in residential homes. But we worked on this systematically – and this is the decisive point – with the whole range of expertise that such an automotive manufacturer has.

We are still in optimization loops when it comes to development of our CHP system. For instance, the suppliers have to learn how to cope with us, and vice versa. The entire project is a process of working toward a goal, which implies that we will not be able to present the optimal CHP system immediately. (Willand, VW)

Entering a market with a product that has just been developed always poses technical and economic risks. VW has implemented a rigid reporting strategy to increase transparency and create a positive perception of the endeavor within the corporation.

The CHP unit is not an innovation by itself, but for the VW plant Salzgitter it is indeed a novelty. There is no history with processing such projects. Hence, we here are not so professionally positioned as would normally be the case in the final development phase of a project when it comes to the commercial dimension. The project is a little softer and very oriented toward research and development.

We have obviously thought about what the CHP system can cost so that LichtBlick can remain within the bounds of what is economically feasible. We monitor our costs and also implemented target costing. Yet we still encounter surprises. But our board has its eyes and ears open. We report regularly on how production is progressing because, alongside those who support the project, there are also others within the business who are rather more skeptical about CHP production. Accordingly, we have to account for what we do. (Willand, VW)

The next step for Volkswagen is to apply its organizational expertise to implement mass production of the micro CHP units.

We already have a few CHP units that operate to our full satisfaction in Hamburg homes. Now we must get down to the routine work. The product must be made better, with fewer costs.

One major challenge on which we are still working is production in large series. Large series processes promise to make a product more economical. A product that we have not yet produced in series has to be made suitable for serial production. For instance, in serial production at Volkswagen it is difficult to imagine spending three hours cabling switch boxes. For us, it must be relatively easy to add elements. We have worker management in our production plant, which means that workers can see on a screen precisely what they have to do, which screw to tighten, and what to plug in. This enables us to guarantee quality assurance measures as we do for our top products, the W12 petrol engine featured in the Phaeton, or Bugatti engines. We have to adapt the production of a CHP system to these assembly techniques to ensure cost-effectiveness. This applies not only for the production team, which has to learn new processes, but also for suppliers.

We are increasing production figures. Currently, production periods are unacceptable for a fully coordinated production process, such as we expect for building cars and engines. We are still a bit slower. Some time will have to pass before we – or our customers – can supply the whole of Germany with CHP units, but we will have the product quality that can be expected from Volkswagen. (Willand, VW)

The company applies its high manufacturing standards also to its cogeneration units.

If the name Volkswagen is on it, Volkswagen must also be in it. It is as simple as that. The customer drives his Golf, his Passat, his Phaeton every day, is generally satisfied with his product. Volkswagen stands for quality. If a CHP unit with a Volkswagen logo on it is in the basement, the customer also expects that the technology will work and will be just as reliable as his Phaeton or Golf. The expectation pattern with Volkswagen is completely different than with smaller firms where people are prepared to put up some imperfections. With Volkswagen, nothing will be tolerated. When we go to market, we want to start with a top-quality product. (Willand, VW)

Figure 2.5 Industrial production of micro CHP units at Volkswagen
Source: Auto-Medienportal/Volkswagen (2012).

VW established the new construction line of micro CHP units with a dedicated project team.

> In the beginning, there was no team but already technical activities and a number of people who had already worked in the field. They were simply uncoordinated – as is usually the case with new processes. For instance, the colleagues in our test facilities, where motors are first built, had competences they had acquired during the manufacturing of the first CHP. But CHP was not automatically included in their job descriptions or written on their doors; rather they were still in their original function.
>
> The first task was to build a project team in the Salzgitter site. This was to include all sub-project leaders, that is, all representatives of specialist areas that need to develop, produce and finally market a CHP (distribution, production, production planning, logistics, etc.). For development, we cooperated inter alia with Ingenieurgesellschaft Auto und Verkehr GmbH, a 50 percent subsidiary of our group. For all dimensions that are needed in such an undertaking, from development through to delivery or distribution, there was a sub-project leader with the competences of their specialist

areas behind them. Depending on the phase of the project, between 30 and 130 persons were temporarily involved.

In the long term, we count on around 160 secure jobs here in Salzgitter thanks to the CHP. (Willand, VW)

The car manufacturer chose its production site in Salzgitter because of the competences in the diversity of motors constructed there.

The Salzgitter facilities are used to cope with a high level of complexity – more than in any other VW manufacturing site in the world. Three-cylinder to sixteen-cylinder motors are produced here, including the Bugatti motor. We also produce industrial motors for a wide range of applications: forklift trucks, street cleaning vehicles, and marine motors. Thus, there is also competence that we send out from here, to other facilities around the world. It was a clear decision to select Salzgitter for, in VW dimensions, a small production series. (Willand, VW)

Additional sources for value creation and maintaining the local workforce played a role, too.

At Volkswagen, and especially at Salzgitter, we search for new business fields in order not to be exclusively dependent on the production of vehicle motors. Typically, we have in Salzgitter development competences in the component area, for instance the motor with its oil pump, with its water pump, with low-pressure pump, with camshaft, etc. All other products go into the actual end product, be it a boat, an auto or a municipal vehicle. We have little experience with systems, for instance autos. This competence needed to be built up. It was a challenge, but helped to secure the location's continued existence. Products that can perhaps be replaced by other makers who can still make them more cheaply need to be substituted by systemic products that are more ambitious and have more content. The CHP is the only finished end-product that leaves this works. That calls for new processes that are typically located at our headquarters in Wolfsburg. (Willand, VW)

For the automobile company, communication with equipment manufacturers of the heating sector proved to be different than what they were used to.

We typically do not build heating units or similar products. Correspondingly, our traditional suppliers do not come from the heating sector. It is always a surprise to see how things are done in other sectors. In the heating sector, also in supplier firms, it is a matter of centimeters, whereas in the automotive sector it is a matter of micrometers or fractions thereof. When we spoke to suppliers and presented our quality requirements, we got astonished reactions. We had difficult and lengthy processes to come together.

The way in which you negotiate with suppliers in the automotive sector is completely different from how you negotiate with suppliers in the heating sector. Many firms were rather frightened and shocked. Those hurdles had to be cleared. The manufacturing of CHP units here at VW in Salzgitter is primarily a technical challenge, but it has many components that are not technical in nature. (Willand, VW)

Volkswagen considers the cooperation with LichtBlick a highly beneficial business model – also for the final customer.

We have linked the LichtBlick philosophy with our product in this business model. In contrast with all other CHP manufacturers, we sell our system exclusively to LichtBlick and have no private customers. Thanks to the concept of swarm intelligence, the quantities that we can sell to LichtBlick cannot be replicated by competitors. We have a win-win situation, even a win-win-win situation – Volkswagen, LichtBlick, and the final customer. (Willand, VW)

Being the first-mover entails the inherent danger of rushing too quickly to the market before other manufacturers and retailers imitate the model.

The plans that we developed jointly with LichtBlick were and are ambitious. Here in the business, we typically have a development process in which we develop products. These processes have a very specific duration, for instance 48 months, in which we then build an auto or a motor. We have taken much less time for the CHP system because we feel a need on the market, our customer feels a need on the market. When you are the first-mover and other competitors in the field are already starting to grasp some of your good ideas – and I believe that there are very good ideas in this concept – then they might possibly be quicker than us. It was therefore clear that we wanted to go public, to make a point and to say: that is our aspiration. Now we just have to deliver. (Willand, VW)

The VW representatives do not see any immediate threat to their market position. However, other car manufacturers may become competitors in the near future because they can apply their expertise to implement proper mass production and achieve the necessary cost degression, too.

At the present time, there are few competitors in the automotive area. If there are cooperative ventures between generator builders and the automotive industry, this will tend to be limited to the supply of motors to a producer. The latter will build them into a CHP unit and sell it for more than we currently project.

A CHP unit constitutes high functionality in a constrained setting. That is typical for car manufacturing. For that reason, I think that another auto maker could certainly become a competitor, whereas firms that assemble CHP plants in, so to say, a larger garage behind their house would not, not least because of the amount of units they are able to produce. A true competitor can only be a business that knows about serial production, that is, automakers like ourselves.

The investments will only pay off if the corresponding numbers can be expected. And you have to be able to produce them. The auto industry is well placed to do this. In principle, an MGM Opel, a Ford, a Daimler could obviously also pursue a similar strategy. (Willand, VW)

Car manufacturer Daimler would consider a cooperation within a consortium for developing micro CHP units, but the market segment may be too small to be attractive for the company.

If we wished to enter – beyond our automotive operations – the CHP market, we would look for partners and not do the research in-house. The right motor would have to be selected from our range, and a dedicated small unit of our firm would then build and test it, similar to our project "Car-to-go." The working group would be very probably taken out of the company for a predefined period of months, a project team would look for cooperation partners, draw up contracts, develop concepts, and then take swift economic action and get down to business. I could only imagine doing this in a consortium. But it is a completely different concept than, for example, investing in wind-powered energy as we already did. At the end of the day, it has to be assessed whether it is economically relevant for any company, in particular in terms of business volumes in relation to the actual turnover that the company has. If it is to be a purely economic undertaking, thought must be given by example to what turnover of €1 billion actually means for a company with a turnover of €100 billion. (Müller, Daimler)

Box 2.3 Small wind turbines – innovations for downsizing in urban settings

Small wind turbines have gained worldwide attention, and market shares are growing – even if the current contribution to energy supply is still minimal. Small wind turbines are generally defined as having 100 kW or less. Global sales amounted to approximately 10,000 units in 2009.

Small wind turbines can be installed on top of buildings in urban settings, or in rural settings where homeowners have enough surface space to privately erect them on their grounds. The American Wind Energy Association calculates that around 15 million homes in the United States would be suitable for small wind turbines, given the size of their land being larger than half an acre and the availability of sufficient wind. If all those homeowners installed a 7.5 kW wind turbine, total generation capacity would amount to 113 GW.

Figure 2.6 Quiet Revolution turbine at Environment Energy Centre, Leyland
Source: Quiet Revolution (2012).

Small wind turbines can be constructed in the conventional shape like larger rotors, but their noise and vibration characteristics may not be suitable in densely populated areas. An innovative solution has been developed by the UK manufacturing company Quiet Revolution. Their qr5 turbine uses a helical, twisted design on a vertical rotating axis, which is easy to integrate with existing masts and buildings. According to the manufacturer, its design virtually eliminates noise and vibration, compared to conventional rotors. This is particularly important because turbines are often installed in direct proximity to residential dwellings. In a typical location near buildings, where the wind is turbulent, they outperform the conventional design, leading to 20 to 40 percent higher energy production. By 2012, the company had sold more than 100 wind turbines in the United Kingdom, the Netherlands, Germany, and Austria.

Regulatory incentives for small wind turbines are available in many markets. In the United States, grants and loans from US Treasury payments and the US Department of Agriculture's Rural Energy for America Program (REAP) funded about 200 small wind installations, totaling 5.8 MW in 30 states. The UK government has established a special feed-in tariff for small wind turbines. According to AWEA (2010), the major barriers for small wind turbines in the USA relate to nature and wildlife preservation guidelines, and local permitting and zoning procedures.

(*Sources*: AWEA, 2010, AWEA, 2012, Quiet Revolution, 2011)

Micro turbines

Aviation technology for the energy market

Micro turbines share the same basic thermal process design with their larger counterparts in combined-cycle gas turbine plants or airplanes: Fuel is mixed with air in the combustion chamber and burns. The combustion gas expands and creates mechanical energy to let the turbine shaft rotate. Micro turbines typically have a single shaft that connects the turbine, the compressor, and the generator. Often they are linked to cogeneration technologies because the remaining heat of the exhaust gases is recuperated.

The predecessors of micro turbines include automotive and truck turbochargers, and auxiliary power units for airplanes and small jet engines. Within the past years, spillover effects from other industries allowed manufacturers to construct micro turbines with significantly smaller sizes, ranging between 25 and 500 kW (WADE, 2012) – enough to provide electricity and heat, if recuperation is activated, for hospitals, hotels, and residential housing units.

The downsizing allowed smaller manufacturers, including Capstone, Turbec, and Ingersoll-Rand, to expand in this niche application. Within the spectrum of decentralized generation technologies, micro turbines are well-suited for providing remote power; they are more robust against temperature fluctuations than internal combustion engines, and highly reliable. They can also be permanently installed at remote sites. The major pitfall of the technology is that some of the current designs suffer in their lifetime from being switched on and off too frequently, and that the overall efficiency is not as high as with other technologies.

One of the pioneers in this niche market is Greenvironment, a European-wide operating technology supplier and system integrator of power stations for decentralized power generation. The initial doubts about the market potential have gradually been dismantled, according to Radu Anghel of Greenvironment.

> When we spoke to market players for the first time at the EuroTier Fair in 2005 and expressed that we wanted to bring gas micro turbines to Germany, the general view was that gas micro turbines had no chance whatsoever on the German biogas market. There are many reasons for this view. The main one is that the electrical energy efficiency for generation using turbines is below that for motors. Each percentage point in the Renewable Energies Act makes a difference of €10,000 per year. Nevertheless, the total cost of ownership is improved by fewer maintenance breaks or longer periods without downtime, that is, the longer service life counterbalances the lower efficiency.
>
> We analyzed our potential customer groups and were clear in our own minds about which markets could be lucrative and which one we should concentrate on. For instance, a gas micro turbine with 500 kW is not interesting if it stands alone in a field, but it is if there is an industrial estate nearby that can be supplied with process heat.
>
> Gas micro turbines are much more respected today because people have seen that the technology can work. Customer attitudes have changed, too, as they are looking at the issue more closely and know what they can demand from us. (Anghel, Greenvironment)

The predominant fuel-intake of micro turbines is fossil energy, but biogas has received some attention, given its carbon neutrality and attractiveness in rural, agriculturally oriented locations – in particular because it would be too costly to transport the raw material across large distances due to its low energy density. In developing countries and countries with a weak transmission grid, micro turbines can ensure grid stability by delivering a more steady electricity supply than other renewable sources like wind and solar, compensating for fluctuations in the energy intake of most other renewable sources.

For decentralized and sustainable energy provision, micro turbines fueled with biogas are a particularly attractive option. Biogas technology was only introduced fairly recently in many markets.

Figure 2.7 shows a Greenvironment micro turbine CHP plant with biogas in German village Muntscha with 400 kW electrical power output – thermal power output of 434 kW; it was commissioned in October 2011.

Biogas technology has delivered several technological breakthroughs over the past decade, and in Germany around 2,800 plants were in operation by 2006, mainly operated on farm sites with gas engines.

Figure 2.7 Micro turbine CHP plant with biogas in Muntscha
Source: Greenvironment (2012).

While micro CHP units may remain financially viable without sustained government support, micro turbines based on biogas combustion may suffer significant drawbacks in their market rollout.

> In the area of biogas, it would be a problem if the subsidies were reduced, since the marginal costs are about 15 cents higher than what can be earned on the market. Hence, it will have a massive influence on the market if subsidies are removed. (Anghel, Greenvironment)

By contrast, the micro turbine market with conventional natural gas would still be economically attractive.

> There would be some movements in the natural gas market, too, but a halt of subsidies would not kill the market. Clearly, the thresholds for market entry would rise, since investments would have to be amortized over a longer time period. (Anghel, Greenvironment)

As opposed to the diverse and fairly large market for micro CHP units, the micro turbine industry is still dominated by only a few medium-sized manufacturers, whose production output is often below 100 turbines per year. Consulting firm Forecast International predicts that the output of one of the largest producers, Turbec Micro turbines, will fluctuate around 70 units per year until 2020.

> Those familiar with the market generally assume that continuous growth of around 50 MW a year is possible in the output class between 50 kW and 2 MW. We estimate that 10 to 20 percent of this market comprises precisely

those niche applications in which we fit gas micro turbines. (Anghel, Greenvironment)

Other companies like manufacturer Viessmann are also considering entering the market for biogas:

> The limiting factor for CHP is always that it needs a consumer who can use the heat. Since this is not the case everywhere, we plan to build a second biogas pilot installation, which will not use the heat immediately but will produce so-called biomethane from the biogas and feed it into the normal gas grid so that it can be consumed elsewhere, like natural gas.
> In 2010 we also presented a business model to install a CHP unit and at the same time offer a contract for the supply of biomethane from any destination. Then the installation can be built where the heat sink is. (Greis, Viessmann)

Market entry and sales strategies (Greenvironment)

As opposed to cogeneration, biogas micro turbines were introduced in the German market only very recently. Despite their apparent environmental benefits and potential for value creation in rural, agriculturally oriented areas, the technology suffers from multiple drawbacks, including a cost structure that largely depends on subsidies and faces public skepticism. Radu Anghel of Greenvironment describes the initial hurdles the company faced when introducing the technology, exacerbated during the height of the financial crisis:

> Since we knew about the concerns and reservations of the market, we started to build, finance, and operate installations as pioneers on the market. We bought the necessary biogas directly from farmers. We started by building four installations. The acquisition of gas from the farmer was regulated in a clear contract. We measured the calorific value of the gas and were then able to buy gas by the kWh. The market for biomethane is now relatively mature, and works on this principle. (Anghel, Greenvironment)

Compared to micro CHP units, micro turbines have technical benefits:

> The gas micro turbine has the advantage that the efficiency level remains the same, independent of very high or very low delivery temperatures. This is not the case with a reciprocal engine, where 50 percent of the thermal capacity is delivery via cooling at 90°C. If there is no use for this temperature, the reciprocal engine has to stop or expensively dump the heat. (Anghel, Greenvironment)

Greenvironment provides guarantees and ensures operation and maintenance for the micro turbines.

> Since the gas micro turbine has a long service life, we can offer 10- to 15-year contracts, which is very interesting for businesses and especially for energy contracting companies. Another unique selling proposition is the high availability, on which we trust by offering warranties for our customers. For instance, with partner company Nordmethan, we guarantee 98 percent availability. That is extremely high for the market. (Anghel, Greenvironment)

Greenvironment's strategy to further establish the company in the biogas business is to begin manufacturing micro turbines and micro CHP units.

> We see good chances for positioning ourselves on the biomethane market by distributing our installations directly via biogas installation-builders and marketing ourselves as builders of gas micro turbine and CHP units. In this way, we can widen our distribution and no longer have to compete directly at the customer level. (Anghel, Greenvironment)

In 2009, the company changed its business strategy to offer micro CHP units in addition to the micro turbines.

> We received financial support for building the first installations from an English investor who made a double-digit million Euro fund available. The concept worked well to start with.
>
> However, as a result of the financial crisis and the very low return on this kind of investment, the fund was terminated. This meant that we had four installations operating without finance. In times of crisis, we could find no banks or investment funds. Sticking to the original strategy proved to be a serious problem because it set the business back. It would have been smart to change the strategy right at the start of the crisis, to sell the installations and to allow financing via the customers. It took almost all of 2009 to work out a new strategy.
>
> As part of our strategic reorientation, we decided to enter an additional market, namely the natural gas market – a classic CHP market. There is the so-called mini-CHP law, which subsidizes CHP units with an output of up to 50 kW. And we distribute a 50 kW turbine, which fits optimally in this market and is very cost-effective.
>
> The main reason for this reorientation was the chance of better development possibilities. For us, the years 2009 and 2010 were a period of focus. We redefined our potential customer groups and looked out for new markets. (Anghel, Greenvironment)

In contrast to the strategy pursued by LichtBlick, Greenvironment does not target individual consumers in the residential sector. The company rather focuses on establishing virtual networks to participate in secondary energy markets.

> Our installations are essentially profitable in the B2B area. In the housing area, there is the problem of the so-called heat sink, that is, an unstable heat pattern with many peaks. Whereas in winter a lot of energy and even an additional boiler are necessary, in summer the demand is very low – hence, the capacity uptake is also very low. This means that the project is simply not profitable.
>
> We want to participate in building virtual grids. This is also a further reason why we do not deal with end customers but operate via contractors. They have the possibility to pool installations. If there are intelligent installations from the outset, we can build virtual power stations, which can participate in the balancing market. At the present time one needs around 15 MW to participate in the balancing market or the minute reserve. As a stand-alone customer, this is not feasible, which is why the concept only makes sense for energy suppliers. Our installations are fitted with a lot of electronics, which makes them easy to adjust and capable to participate in the minute reserve.
>
> We want to build up local intelligence so that the installation is the central point of decentralized energy production. It is locally decided what happens; that is, if it is necessary to switch off the unit due to participation in the balancing market, it must be ensured that the customer does not notice anything in his heat supply, and has sufficient buffered heat available in his storage device.
>
> B2C business is much more complicated than B2B business since final customers have very different expectations. By contrast, contractors' expectations can be forecast much more easily. We therefore limit ourselves to the latter category of business. (Anghel, Greenvironment)

In the market for micro CHP units, Greenvironment faces tougher competition than with the micro biogas turbines. One competitive advantage of the company is its expertise in process control.

> All CHP unit builders are essentially our competitors. There are many, including consulting engineers, who get hold of a motor and convert it into a biogas motor. This is not a great challenge if you have some technical expertise.
>
> At the moment, we are in the mid-field of the CHP market, but would like to become a market leader in this field. To reach this goal, we differentiate

ourselves from our competitors, firstly through technology. In the area of gas micro turbines, we are the only one, which makes it easier for us to open doors. Moreover, we have worked on the theme of installation control. Although we are not the only one on the market, we do have a promising product with the gas micro turbine and can even carry out remote maintenance operations on our installations. This differentiates us clearly from our competitors.

In the case of micro turbines, competitors would be welcome and would even help us in our present situation, since more references increase the profile of the product and convince customers that the technology is viable. (Anghel, Greenvironment)

According to Anghel, the threat of market entry of the big electricity companies seems fairly low. However, they will most likely remain the dominant players in the overall energy market in the near future because of the capital-intensive nature of energy supply.

Market entry is not interesting for the big players, for several reasons. A great deal of know-how would have to be built up. In addition, the whole energy market currently faces a difficult situation. The nuclear reactor debate is only one part of the spectrum.

A trend can be seen in the need to create a more flexible market that reacts more rapidly to systems and network fluctuations. This poses a major challenge for network operators and energy suppliers. Our products offer flexible energy solutions here, which is why we see them rather as customers and not so much as competitors.

There are trends that suggest that the market will further fragment. Even the big four German electricity companies are currently having a hard time. But ultimately they have the money, which is why I do not think that they will disappear from the market. They will certainly manage to adapt to the new structures and to react more flexibly. Even now we can see that smaller units are being created and efforts are being made to look at customer needs more rapidly.

The energy market is very capital-intensive. Competitors will find it difficult to enter the market and establish themselves. Industrial undertakings in particular will look for security, since they cannot afford to do without energy for a given period just because they have made the wrong choice of energy business. They will tend to stick with traditional business models.

I do not believe that the structure will change radically, though. The challenge lies in becoming more flexible and networks also have to be expanded. Electricity must be delivered where it is needed. But the large offshore production facilities, for instance, are situated where electricity is not needed. And only the big players can build the necessary infrastructure.

They also need someone to take responsibility for the network. Electricity is not something that can be stored. Only strong businesses can assume responsibility for the grid. Unfortunately, there is not much money for network expansion just now. Given the political situation, subsidy issues, etc., it is difficult to expand a network and then clarify with the Federal Network Agency how high the transmission fees must be. Structuring this in a very transparent way is very difficult. (Anghel, Greenvironment)

Cooperations and alliances (Greenvironment and Viessmann)

Alliances between manufacturers and service companies can be observed quite frequently in the energy sector's niche markets. For example, Greenvironment relies on a cooperation with Capstone Turbine Corporation, a US American gas turbine manufacturer founded in 1988 that specializes in micro turbines, and the Italian manufacturer Articomp. According to Anghel, the cooperation is mutually beneficial for the respective companies because it gives them planning security and feedback for their product development.

> We continue to benefit from a cooperative venture with the Capstone business, which has existed since the firm was created. Capstone placed a 200 kW turbine with a high efficiency on the market. With orders for this turbine type, we gave Capstone the necessary planning certainty for development of the turbine. The high level of efficiency opened up new segments for Greenvironment in both the biogas market and the natural gas market. At the end of 2009, contracts were signed for two biogas installations as well as a large 600 kW natural gas installation.
>
> We also cooperate with the Italian firm Adicomp to advance the development of gas compressors, which are able to deal with different gas qualities. We give our know-how to Adicomp so that they can develop a new system by using this know-how and from which we can in turn benefit.
>
> In other areas, we work correspondingly with other businesses. We buy all the necessary components from the manufacturers and have them assembled by a specialized supplier in Saxony. Then we sell the aggregated unit as a plug-and-play plant to our customers. (Anghel, Greenvironment)

Greenvironment also enters joint ventures with municipal utilities.

> In early 2010 we implemented another new strategy for the first time – a joint venture with Schmalkalden's municipal utility, known as Decentralized

Energy Systems Schmalkalden. The idea behind this is to sell service and operate a CHP unit in Thuringia. This joint venture expanded at the end of 2011 with the entry of a new subsidiary of German utility RWE, RWE Energiedienstleistungen, as the fourth partner.

In the case of Schmalkalden's municipal utility, our involvement has a longer-term strategic background since smaller municipal utilities, just like their larger counterparts, also have larger commercial or industrial customers in their portfolio. Since these are lucrative, there is stiff competition for them on the market. By buying additional gas, smaller municipal utilities can keep these large customers. This creates a win-win situation for both the municipal utility and for us. As opposed to our parent company, we see ourselves as an installation provider. We offer our service at a certain price, while our parent company understands it as a more holistic strategy.

In the natural gas market, we are also working on further cooperation in joint ventures, targeted closely on the high-temperature market. For instance, we see chances with pharmaceutical and dairy businesses. Like many other companies, we also work closely with other SMEs. The idea behind this is to use synergies, since each partner brings in its know-how for the generation of new ideas. (Anghel, Greenvironment)

Another player in the market for biogas is the family-owned company Viessmann, an industrial manufacturer based in the German state of Hesse. The company uses its expertise with cogeneration to conduct experiments with new biogas plants. Manfred Greis, in charge of corporate communication and sustainability at the company, explains the advantages of biogas compared to other renewable energies like wind and sun:

As part of the political double strategy, we would like to enhance efficiency by consuming less energy but also by generating more efficiently via combined heat and power. Since the national and global primary energy supply has relatively small contributions from solar energy or geothermal energy, it is clear that biomass will play a large role here. Biomass has the advantage that it does not fluctuate. In fact, it is really stored solar energy. It can be used during the whole year. And it can be made available as a domestic energy source; it does not have to be transported from somewhere. For our company, it was also clear that we did not want to import pellets from overseas, which would then have to be carried across the world's oceans. Rather, our aspiration with regard to biomass is that we want to move along the line of sustainability – also, in order not to get involved in the food-versus-fuel discussion.

> Thus, we built a biogas installation, a first one, which works on the principle of dry fermentation. It uses mainly organic material from agriculture as a substrate, that is, natural waste and lop from protected conservation areas. Nobody eats this. At best it is composted. Here we can produce biogas with it. We have a CHP unit and we produce heat and power with these substrates. This is easily possible for us because we have built it at an industrial location in which we have the corresponding heat sink so that we can use the heat. (Greis, Viessmann)

The company is also pursuing the technological option to chemically transform the biogas into bio methane, with the possibility of feeding it into the central grid. This strategy eases the burden of finding an appropriate consumer of the heat supply.

To explore the potential of biogas, Viessmann closely cooperated with local farmers when it established its first plant, and integrated them into the learning process.

> Since we are at home in the rural part of northern Hesse, we have started to buy agricultural land around our location, in particular stretches that are difficult or impossible to use for regular farming, 170 hectares in total, and have planted poplar trees. One of our aims was also to learn about the entire process chain, which we want to develop further and generate know-how that we can pass on to our target groups. In this project, we have worked very closely with the local agriculture sector, which we need to take with us. Otherwise, they would ask what this manufacturer intends to do with agricultural land and whether we would start competing with them. On the contrary, we actually wanted to work together. We started a cooperation with an environmental NGO, the Nature And Biodiversity Conservation Union, too, so that we can incorporate the nature protection aspects and prevent resistance from that side. We took advice on what to plant along the edges, how to make corridors for the fauna. These local associations have, for instance, set up watching posts so that birds of prey can hunt mice more easily. (Greis, Viessmann)

Viessmann traditionally co-operates with local technicians and craftsmen as their partners. However, if the plants exceed a certain size, local craftsmen may not have the sufficient knowledge to cope with the challenges, and Viessmann takes over the responsibility.

> We have traditionally been organized in such a way that our market partners are artisan firms. We see it this way in the future, too, because we need a qualified local partner for the investor and the end user. I believe that the

industry cannot do this and that the structure of an energy supplier cannot do this, at least in the foreseeable future. We have to qualify local technicians, have to assign them a role also in the future. On the other side, we are clearly moving into new areas that did not play an important role for us in the past, be it biomass incineration installations or local heat networks. This is obviously too much for typical artisan firms. For that reason, we as an industrial undertaking develop our own expertise in these fields. (Greis, Viessmann)

A cooperation with large energy utilities like E.ON also provides mutual benefits.

We would like to offer integrated solutions. We are already talking to and cooperating with energy suppliers. For instance, our new micro CHP is also promoted by E.ON. Those who deploy this machine will receive support from E.ON because the company is equally interested in bringing these technologies to market. There are already cooperative ventures. As electricity and heat have grown together, structures are also converging in terms of suppliers and manufacturers. This will obviously bring change. Models such as those used by Volkswagen and LichtBlick will certainly play a role in that transformation. (Greis, Viessmann)

Findings on micro CHPs and micro turbines

- *LichtBlick's business model shows that micro CHP units are commercially viable on the residential level*: While the predominant business strategy in the deployment of micro CHP units still relies on a heat demand that ideally exceeds 5,000 to 6,000 hours per year, new market players like LichtBlick successfully develop alternative operating modes that ensure profitability even in residential settings with a significantly lower heat intake. With its swarm electricity concept, the company bundles micro CHP units in a virtual power plant and uses peak prices on the electricity wholesale market to ensure profitability. LichtBlick demonstrates the potential for innovative solutions that exploit opportunities created on the basis of the liberalized market. The imitation of the LichtBlick strategy by Vattenfall and Deutsche Telekom indicates the potential of micro CHP units in residential settings.
- *With mass production launched by Volkswagen, micro CHP units will become a major application in a decentralized energy supply*: Micro CHP units are still predominantly produced in small series and largely manual manufacturing processes by small and medium enterprises. With car manufacturer Volkswagen, the first real prospect of industrial mass production is in sight.

> *Box 2.4* **Energy poverty, village level entrepreneurs, and the Nuru Power Cycle**
>
> In rural Africa, India and many parts of the developing world, poor families spend a substantial share of their income on energy, mainly kerosene, batteries or charcoal. Even if the long-term government objective is full electrification, remote villages may get connected in a distant future, or not at all. Even island systems powered by Diesel generators or micro hydro may be beyond the reach of local residents.
>
> With the success of cell phones even in rural areas – not only for communication, but also, for example, money transfers – electricity becomes an essential necessity of everyday life. Similarly, lighting may be provided with conventional resources like kerosene. However, kerosene is harmful because its fumes may cause respiratory and eye problems, and can cause fire incidents. By contrast, the cheapest solar lighting products range from US-$10 to 25 and are under regular circumstances too expensive for customers earning just US-$1 to 2 per day. In addition, solar-powered devices need sunlight, which is unavailable during the night, on cloudy days, or often during rainy seasons. Solar lamps left unattended to charge are also susceptible to theft.
>
> Sameer Hajee, the founder and CEO of a small enterprise called Nuru Energy, and his team observed that 90 per cent of the needs for lighting after dark were task-based "from milking cows to doing homework and cooking. They don't really need to light a whole room most of the time." Hajee developed a solution with a one-of-a-kind robust and simple-to-use off-grid recharging platform, a pedal generator called Nuru Power Cycle. The mini generator can be used to recharge portable modular LED lights, in total five at a time in 20 minutes, with each light providing up to 26 hours of light to the owner. In addition, the cycles can be used to recharge mobile phones and radios.
>
> To promote his product, Hajee set up his own network of rural entrepreneurs who can sell or rent the lights and offer a recharging service. His company also acts as a coordinator between microfinance institutions and so-called village level entrepreneurs. With recharge fees around US-$ 0.20 those entrepreneurs are able to earn up to US-$3 per hour. "Given that many of these people earn less than US-$2 per day the income revenue is enormous. It is the same cost as a phone recharge in the country."
>
>
>
> *Figure 2.8* Recharging with the Nuru Power Cycle
> Source: Nuru Energy (2012).

(Carrick 2009) For the owner of the LED device it costs about one-sixth of the price of kerosene to recharge. Each POWERCycle equipped village level entrepreneur is able to support up to 500 Lights, which services up to 2,500 people.

To date, Nuru Energy has set up 70 village-level entrepreneurs who have sold around 10,000 Nuru lights. For its efforts, Nuru Energy has been recognized as the recipient of numerous global awards, including the 2010 UNDP World Business and Development Award. Its founder, Sameer Hajee, who has Canadian citizenship with roots in Kenya, was nominated a "Social Entrepreneur of the Year in Africa" by the Schwab Foundation for Social Entrepreneurship in 2012. His company operates in several sub-Saharan African countries and in India.

(*Sources*: Nuru Energy, 2012, Carrick, 2009)

Given the expertise of the company in terms of automation and rationalization, chances are high that micro CHP units will satisfy requirements of a mass market product once VW moves to higher output numbers.

- *Micro-generation with biofuels will only commercially survive in a carbon-restrained policy setting*: Micro-generation based on biofuels is still not cost-competitive and depends on appropriate government incentives to reduce investment uncertainty and provide an adequate cash flow. The use in agricultural niche applications may be hampered if subsidies stop. Given the long-term policy objective of substantially reducing greenhouse gas emissions and internalizing environmental externalities from fossil fuel combustion, the reliance on a domestically available energy source, in addition to the positive employment effects in rural socio-economic settings, sustained state support for the technology is likely.
- *In-house expertise in information technology and legal matters will become a core asset in the liberalized market*: LichtBlick's organizational structure shows that sustained success in liberalized energy markets can most likely be achieved if competences to handle regulatory questions, expertise on available niches, and increasing informational complexity are bundled inside the company.
- *Innovation networks provide the base for corporate survival in an early stage of technological deployment*: A close feedback among manufacturers, retailers, and operators of technically less-advanced technologies – like micro-generation based on biogas turbines – creates synergies by reducing financial risks for the manufacturer and ensuring quality improvements for the retailer and operator. Innovation networks and joint ventures between municipal utilities and operators can enrich the utilities' generation portfolio by high-value, environmentally sound supply structures. The integration of stakeholders from the agricultural sector and non-governmental organizations into the planning and operating process fosters acceptance of the new technologies.

3
The Rise of Island Systems

Decentralized energy generation allows consumers to become producers. The motivation is, of course, in part related to financial rewards, but it also offers a value proposition beyond monetary aspects. Citizens can regain a degree of autonomy and self-determination over one of their essential needs – energy. During most of the 20th century, the policy of public service obligation guaranteed a fully functional and reasonable, but passive reception of energy services for every household. Similar to the manifold possibilities, the internet offers individuals to become active in wikis, blogs, and social networks, thereby giving them a wider range of capabilities with the freedom to choose and participate.

The integration of former consumers into a process of collectively defining virtual realities is paralleled by local, real movements like bioenergy villages, which define the sources of their energy supply by communal processes and a bottom–up stakeholder discourse.

Municipal and regional utilities also discover consumer empowerment and local participation as a key marketing and selling opportunity of their services. The proximity to the social context of their customers and the trustworthiness they convey helps them to establish new sources of value creation in rural as well as urban areas.

Theoretical framing

Capabilities and collective empowerment

Any individual's state of well-being not only depends on his or her mental predisposition and preferences, but also on the contextual – economic, social, even geographical – conditions to achieve happiness in life. Noble prize laureate Amartya Sen calls these "capabilities," for example the ability to have access to a proper health and education system, or the ability to participate in communal decision processes, from which individuals then have the freedom to choose: "There may indeed be a case for taking incomes, or commodity bundles, or resources more generally, to be of direct interest in judging a person's advantage, and this may be so for various reasons – not merely for the mental states they may help to generate" (2003, 191).

A decentralized energy supply enriches the set of capabilities of the individual by offering an additional dimension of freedom or a valuable option they are able to choose. The same rationale can be applied on the communal level. The capability to collectively select the desired form of energy supply represents a new dimension of empowerment: "Empowerment is a multi-dimensional social process that helps people gain control over their own lives. It is a process that fosters power (that is, the capacity to implement) in people, for use in their own lives, their communities, and in their society, by acting on issues that they define as important" (Page and Czuba, 1999).

Empowerment is likely to be a key trigger for why a decentralized energy supply will achieve much higher penetration rates than a mere cost–benefit analysis would suggest. While the terminology broadly refers to the expansion of freedom of choice and action,[1] the World Bank more concretely identifies four key elements that characterize empowerment:

The first element, *access to information*, is considered by the new generation of "digital natives" an essential part of their quality and style of life. They have the skills to navigate and differentiate between useful and useless information on the internet. Role model communities and best practices can be transmitted and communicated more easily, individual households can circumvent information bottlenecks, for example by obtaining independent and neutral advice for building efficiency via online portals like co2online.[2]

When central governments hesitate and postpone policy action to promote the move toward a more sustainable society, progressive communities step in and implement measures on the local level. The erosion of the nation-state in supranational political contexts like the EU comes as top–down movement – shifting political responsibility to authorities like the European Commission or the European Parliament – but also as a bottom–up grassroots revolution to strengthen subnational linguistic, ethnic, and cultural particularities and their respective populations. The increasing autonomy of regions like the Basque country, Catalonia, Scotland and Wales is paralleled by a growing amount of opportunities to intervene in the local political process through referenda and direct democratic mechanisms.

Inclusion and *participation* are the second key elements of empowerment. As the World Bank writes, "in order to sustain inclusion and informed participation, it is usually necessary to change rules and processes to create space for people to debate issues, participate in local and national priority setting and budget formation, and access basic and financial services" (Narayan-Parker,

2002). Whereas in countries like Germany only 1 to 2 percent of the population participate in decision processes, in Latin American communities, the percentage is upwards of 30 (Bewarder, 2011). The rise of the digital society greatly facilitates and eases participatory processes. Citizens can submit suggestions, for example where and how to allocate free municipal funds, via the internet. According to the United Nations E-Government Survey 2012, the most advanced countries in that respect are South Korea, the Netherlands, the United Kingdom and Denmark.

Empowerment on the communal level cannot work without *local organizational capacity*. Individuals often trigger a change process in the collective consciousness, but they need a critical mass of followers who mobilize hesitant individuals within the community and show some experience and professionalism to channel and unify public interest and needs.

The fourth precondition for empowerment concerns *accountability*. The World Bank demands establishing mechanisms both inside and outside organizations, such as administrative and political entities as well as private firms. Measures to increase accountability on the local level include a handing-over of responsibilities from the state to the citizens: "Accountability for public resources at all levels can also be ensured through transparent fiscal management and by offering users choice in services. At the community level, for example, this includes giving poor groups choice and the funds to purchase technical assistance from any provider rather than requiring them to accept technical assistance provided by government" (Narayan-Parker, 2002, 21).

Race to the top and strategic differentiation

In the age of globalization, governments attempt to attract international corporations with favorable conditions like low tax rates or special economic zones like Shenzhen in China. If municipal authorities have some discretion over communal taxation, they may also compete with their peers by lowering the tax burden for investors. In the political economy literature, this phenomenon is commonly referred to as the *Delaware effect* because the US state of Delaware has succeeded in attracting most incorporations of publicly traded companies in the United States due to its attractive corporate laws (Bebchuk, 1992, Cary, 1974). Especially in the regulation of environmental protection and labor laws, the theory predicts a gradual erosion of stricter standards because of competitive pressures.

If the "race to the bottom" were considered the exclusive driver of government policy, a decentralized energy supply would have difficulty gaining support among decision-makers. Evidence across the entire political hierarchy – from municipalities to global organizations like the United

Nations – suggests that a contradictory mechanism counteracts the race to the bottom. Analyzing California's pioneering role in implementing stricter environmental laws than other states, Vogel (1997) calls it the "race to the top," or the "California Effect." He observes that coalitions between public interest groups and local firms – often with the support of non-governmental organizations that pursue their own value-driven agendas – may push politicians to implement stricter standards, even at the risk of deterring some potential investors. The race to the top can be observed, for example, in an increasing number of countries and states adopting costly subsidy schemes or quota systems to increase the share of renewable energies (see Weinmann, 2007, for a discussion). Municipalities seeking voter support – especially following incidents revealing structural problems in the system such as major blackouts or accidents at a power plant – may be motivated to bandwagon and imitate the California Effect and establish an energy system that is not exclusively based on cost advantages.

But it can also be a strategy of differentiation pursued by market agents. Porter (1980) distinguishes three generic strategies: cost leadership, differentiation, and focus. Following a strategy of cost leadership, a company aims to outperform competitors not by offering better products but by offering similar products at lower costs. These might be due to economies of scale, proprietary technology, or exclusive access to raw materials. Pursuing a strategy of differentiation focuses on offering products and services that differentiate the company from its competitors in the eyes of the buyer. Differentiation attributes might be tangibles, such as offering additional services, or intangibles such as a brand name that associates a kind of well-being or belonging to a certain group of people. The strategy of focus emphasizes the idea of not covering all market segments but rather specializing on one or more selected market segments.

Municipal utilities often have different cost and revenue reallocation schemes than companies that operate only to maximize shareholder value. Hence, they have to compensate for their relatively high prices with soft factors that convey the environmental awareness and emotional identification of the customers with the product.

Their value proposition of a decentralized energy supply closely draws on the notion of the race to the top and the strategy of differentiation: Locally produced, environmentally friendly energy is worth more to specific customer segments than finding the provider with the lowest prices. It allows municipal utilities to avoid cost competition and to follow a strategy of differentiation for the regional market or specific target groups. From a broader perspective, it may provide the opportunity to attract new inhabitants to the region, be it due to local technologies and artisan services that need to be provided, or due to belonging to a group of energy independents.

Bioenergy villages

The countryside strikes back

Bioenergy villages are local initiatives that try to achieve a high level of energy autarky by means of local energy resources, in particular biomass. They can be considered part of a wider movement to focus on regional strengths and values, initiated by attempts to decrease the consumption of long-distance food and promote organic, locally grown produce and regional food brands by health- and environmentally conscious consumers.

> Regional origin becomes increasingly important for the consumer, as the food industry shows. This trend may also spread to the energy sector and embed us further in the local community. (Liedtke, SWK)

Bioenergy villages aim to achieve 100 percent energy autarky. While heating is generally covered by the local bioenergy plant, most of the villages still rely on the central grid for electricity provision. Figure 3.1 shows an exemplary configuration of the energy supply of a bioenergy village.

The bioenergy village depicted in the figure consists of three major components: a cogeneration plant based on biogas, a wood chip heating plant, and a local heat distribution network. For peak heat demand, a conventional, fuel-oil-based boiler is added. Electricity is fed into the local distribution network and then transmitted to the local customers. Surplus electricity can be sold to the local utility.

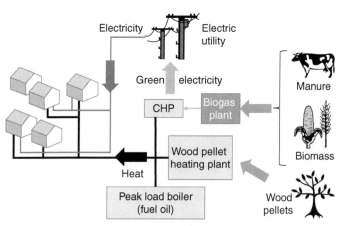

Figure 3.1 Energy-supply concept of the bioenergy village Jühnde
Source: Bioenergy village Jühnde (2012).

In Germany, more than 90 bioenergy villages have been established. But the market potential for bioenergy villages remains closely tied to adequate geographical and agricultural conditions.

Frank Hose, director of regional utility Ostwürttemberg Donau Ries AG (ODR), a subsidiary of German incumbent EnBW, describes the evolution of renewable energies in his company's concession area.

> The ODR supplies a rural area with a great deal of agriculture like corn and other renewable raw materials. At a very early stage, we became a model region and a lot of farmers generated biogas with bacteria in biomass plants, which is used in combined heat and power plants to generate renewable electricity and heat. So today we have a large number of biogas plants and more than 21,000 photovoltaic installations in our region because the sun shines here for more than 1,000 hours a year. Together with wind power, probably 80 to 85 percent of the electricity generated could actually stem from local renewable energy sources by the year 2020. (Hose, ODR)

While urban agglomerations with high population densities and corresponding energy consumption levels will in most cases not be suited for bioenergy – apart from organic waste plants – many stretches of rural areas may produce and harvest the necessary biomass without intervening too drastically in the traditional use of plants for nutrition, according to Alexander Voigt, CEO of island solutions provider Younicos:

> In areas of low population density with a lot of woodland, villages can become self-sufficient simply from waste wood and the exploitation of local biomass. Not even photovoltaic energy may be needed there. Looking at total energy supply, biomass will not be a solution except for individual regions where it may well provide 100 percent of the energy needs and more. (Voigt, Younicos)

Ownership and participatory processes (Jühnde)

The bioenergy village of Jühnde in the northwestern state of Lower Saxony was the first village in Germany to turn to local energy supply. Since the project was launched in 2001, the configuration of its power and heat supply has become the blueprint for many other villages.

In the village, a CHP plant with 700 kW electric power is fuelled by biogas produced out of manure and biomass. For cold winter periods, an additional woodchips heating system with 550 kW thermal power as well as a peak load boiler run by fuel oil supplies the village with heat and warm water. Three

Box 3.1 **Samsø – energy autonomy in the Baltic Sea**

One of the prime examples of a successful transition from fossil to renewable energy in an island setting is the Danish island of Samsø. Within the relatively short period of less than 10 years, the island's approximately 4,000 inhabitants, dispersed over 22 villages, switched to a completely autonomous energy supply. One community has been producing heat by burning straw since the early 1990s; but the decisive change came in 1997 when the Danish government awarded the community "Denmark's Renewable Energy Island." In 2000, there were 11 wind turbines erected on-shore. They are owned either by local farmers or by so-called guilds formed by the residents of the island. By 2003, 10 offshore wind turbines were built, also predominantly owned by locals and municipalities. Meanwhile, three further district heating plants based on biomass have been established. One of them uses additional solar thermal energy.

While in 1997 only 13 percent of the islanders' energy consumption was based on renewables, in 2003 – after the integration of the offshore windmills – full energy autonomy was reached, and electricity even began to be exported to the mainland.

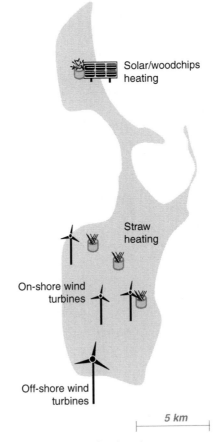

Figure 3.2 Samsø's main energy production sites
Source: Samsø Energy Agency/www.energiakademiet.dk (2012).

> One of the main reasons for the success of Samsø's energy transition was the involvement of residents in the movement. Søren Hermansen, a high school environmental studies teacher and Samsø native, became the first person to receive funding through the Renewable Energy Island reward. He gradually convinced the islanders to shift their energy inputs. He developed the investment plans, whereby local residents could participate in building the wind turbines and solar panels. In 2008, he was named one of the "Heroes of the Environment" by Time Magazine, and one year later he received the Gothenburg Award, the Nobel prize of the environment.
>
> The communities participated in discussions and formed the guilds. Around 300 households installed solar thermal water heating systems, and overall heat consumption decreased by 10 percent due to investment in the insulation of houses and changed heating habits. However, some of the projects did not succeed. For example, the introduction of three electric vehicles in 1999 turned out to be difficult, because the technology was not yet sufficiently advanced, and driving habits did not match.
>
> Samsø has become a showcase, with eco-tourism booming and a number of international cooperations have developed, including the coordination of renewable energy policies and strategies among three European insular regions including the Canary Islands (Spain) and Crete (Greece). A current project of the Samsø Energy Agency is to approach and support households in choosing the most energy efficient products and encouraging them to reduce household consumption in gas and electricity.
>
> (*Sources*: Andre, 2010, Droege, 2009, Samsø Energy Agency, 2012)

quarters of the biomass are harvested by the villagers, the remainder is purchased regionally. When heating demand is low, wood-chips or log-wood are dried with the available excess heat. The villagers were able to reduce their individual carbon footprints from almost 10 tons per year, which is the German average, to about 2 tons annually (REGBIE+, 2009).

Biogas technology is still in the first rollout phase and has not yet reached its full technical potential. One of the reasons, according to the director of the local initiative in Jühnde, Eckhardt Fangmeier, is the lack of experience with biogas technology:

> In the project design phase, we tried to avoid excessive investments while ensuring a high quality of the plant components. After having been in operation for five years, we have now come to realize that our technology was not sufficiently developed. An industrial standard does not yet exist. We are rather in the development phase, which is marked by risks and instability. Even if in 2011 there were around 5,000 biogas plants, the technology is not yet mature enough to be considered a mass product. A certain lifetime or reliability, such as we have with cars, for example, cannot be guaranteed. (Fangmeier, Jühnde)

Fangmeier estimates that it will take some more time until the technology's full technical potential is realized and overall plant efficiency reaches acceptable levels.

> Today we have achieved a very different standard of technology compared to five years ago. On the technical side, it will perhaps take another five years to achieve adequate safety and reliability, though. At present, various methods are being tested to find adequate mechanisms for preparing the biomass before the material enters the fermenter. The process is still highly energy-intensive. The goal is to raise the yield and achieve a substantial reduction in electricity consumption. Jühnde requires 11,000 tons of biomass annually, an incredible amount that passes through the plant. There is a huge incentive to become more efficient. (Fangmeier, Jühnde)

For the construction of the different system components of the bioenergy supply, Jühnde employed local craftsmen. The biogas plant and the CHP plant were purchased from supra-regional manufacturers.

> We outsourced the tenders to consultant engineers. The civil engineering work was awarded to firms in the region; the biogas plant was constructed by a specialized firm located in Neumünster, while the combined heat and power plant was purchased from a big German manufacturer. (Fangmeier, Jühnde)

The power production from the plants even exceeds local demand and can be fed into the central grid.

> I am proud that our CHP generated about 5 million kWh of electricity per year and the villagers consume only 2 million kWh in Jühnde. The supply of heat for the village is also assured. At present 70 percent of the villagers are connected to the heat supply. Interest has not ceased – we are constantly connecting more villagers to our grid. (Fangmeier, Jühnde)

Without state funding, the financial survival of a municipal project like Jühnde could not be guaranteed, though.

> Bioenergy villages are closely tied to the regulation on feed-in tariffs and subsidies for renewable energies. Without this type of incentive, it is practically impossible to realize such an endeavor. We hope that legislators see the benefits of the projects and will continue to support them, for example by providing remuneration when not only electricity but also heat is generated as a co-product. (Fangmeier, Jühnde)

The transformation of a municipality into a bioenergy village requires dealing with a set of complex technical, legal, and economic challenges. Close cooperation among the initiators of the project is essential.

> Our bioenergy village was a pilot project, and we applied for official assistance. When we received notification of support, we founded the cooperative in autumn 2004 and invested in a transformer for the plant. We had a competent consultant engineer who worked on the project with just as much heart and soul as I did myself. That made both technical and economic coordination possible and even allowed us to cope well with the complicated building regulations.
> Within nine months we set up the local heating network and the plant. We began supplying heat in 2005 right on the birthday of the then mayor, which helped in attracting a lot of interest from the media and the general public. We were able to rely on the support of everyone, the whole village community as well as the farmers. (Fangmeier, Jühnde)

Academics from the nearby University of Göttingen were involved in the initial stages of the project. They were valuable and economically neutral advisors in the process. However, the community process gradually dissociated from the scientific support and became self-sustaining.

> Universities and scientists are neutral, unbiased partners without commercial interests, which is very beneficial for project development and for convincing the villagers. We cooperated closely with the university until we founded our cooperative. Then we took over independently. The university by and large opted out of implementation, construction, and operation of the plant, but completed a couple of subsequent research activities with us. (Fangmeier, Jühnde)

Bioenergy village Jühnde received a substantial direct subsidy and a number of loans and grants from several government agencies.

> In order to raise the investment of €5.2 million for the project, we received €1.3 million from the German Agency for Renewable Resources and financed the rest via loans from the German development bank KfW, other banks, and private sources. We are now paying off those loans. (Fangmeier, Jühnde)

In line with the participatory character of the project, part of the money was raised from stakeholders and final consumers, similar to a co-op or association where all members become co-owners of the property. The municipal administration participated financially, too.

> For financing we raised €500,000 of equity capital. Every customer who wanted to get connected to our heat supply had to join the cooperative by purchasing three shares costing €500 each. With 150 heating customers that amounted to €225,000. Some very supportive customers invested even more. The local authority also got involved by reinvesting the money from the sale of the land. In this way, we were able to raise the required amount of equity capital. (Fangmeier, Jühnde)

The longer-term objective is to financially reward the members of the association.

> The members of our cooperative received the first return in 2010 with nearly 2 percent of their investment. The law on cooperatives foresees that, after putting aside certain reserves, we can also pay out dividends. It is my personal goal that the members who have paid us money will at some point receive interest and that the investment they have made will pay off. We are not yet there, but at least the customers benefit from a favorable price for heating. They currently save €1,000 annually, compared to a conventional oil heating system. (Fangmeier, Jühnde)

While prices for fossil energy sources fluctuate – especially natural gas and oil for residential heating – the final consumers in Jühnde benefit from a stable price scheme.

> With respect to energy prices, we compiled a forecast in which we calculated the cost along with the oil price of 35 cents per liter of heating oil, as it was at the point of our construction works. The depreciation of the plant, maintenance, repairs, fixed and variable costs were also factored in. We managed to explain the calculations to the inhabitants in a plausible way and they agreed to the price. To this day we have been able to stick to the same price so that by now the heating customers have gained a financial advantage. (Fangmeier, Jühnde)

Bioenergy villages promote a self-sustaining energy supply – a vision that directly cuts into the current business model of supra-regional utilities. However, instead of confrontation, Jühnde pursued a strategy of cooperation with the large energy supply company E.ON:

> At the start of our project, before we had founded the civil-law partnership, the utility E.ON offered us a comprehensive, all-inclusive package for our

village. But we identified so strongly with the project that we wanted to implement it on our own.

We have no problems with the major energy utilities. We made a conscious decision to buy our transformer from the incumbent's affiliate E.ON Mitte. In this way we are also able to raise acceptance for decentralized projects. We have good contacts with the colleagues from the grid department at E.ON.

We also cooperate directly with E.ON and organize joint information events. In the nearby town of Hardegsen, E.ON has a biogas plant to which we send some of our visitor groups. One of our objectives is to demonstrate to our visitors that there are also other possibilities than the one we chose. Ultimately, they have to decide for themselves what source of energy they would like to use. (Fangmeier, Jühnde)

Local initiatives heavily depend on the continued involvement and agency of individuals. The intrinsic motivation is derived from benefits other than financial gains.

Since I am primarily employed with another firm, the bioenergy supply is what you could call a hobby, but a very intensive one, and challenging for the family and for one's own schedule. Thanks to the internet, my colleagues, and my CEO at work, I can cope, though. The plant manager can ring me at any time, even when I am at work, if something has happened, or if decisions have to be taken. In the evening I drop in at the plant to take a look myself. That helps us to economize on personnel costs. We are running the project primarily as a matter of conviction and not for financial reasons. But if it should happen to yield profits some day, everybody will be able to benefit from it. (Fangmeier, Jühnde)

Jühnde employs one operating manager to run the biogas plant. All other functions are taken over by volunteers or local technicians.

The plant manager has a full-time job and is in charge of operation and maintenance; he also receives the deliveries of biomass. Another member of the board is a farmer and is more easily available during the day than I am. He looks after the biogas plant and takes some part in the work of the operating business. For example, he takes on weekend duties and the emergency telephone. He supports the project on the practical side, whereas I can contribute with my experience in accounting and finance.

Village craftsmen also support us. For example, a local electrician looks after the heating infrastructure. We remunerate him by the hours he spends on working for our project. (Fangmeier, Jühnde)

The provision of local, decentralized energy is a vehicle to enhance empowerment of the residents.

> The involvement of the local population is a central aspect of our project and, more generally, a common feature of bioenergy villages. Technology alone is not sufficient. We let people become a part of the project and give them a degree of supply security. With their money, they participate in the project and take on responsibility. The cooperative is also a community, which aims to continue implementing projects together in the future. (Fangmeier, Jühnde)

For example, the Jühnde villagers plan to integrate electric vehicles into their strategy.

> Electro-mobility will be a key issue of the future. We aim to apply for funding from the political and business community so that we can start another pilot project with electric vehicles in a rural setting. Then we could offer everything: electricity and light, heat for heating space and water, as well as mobility. According to my calculations, with twice the amount of electricity we would be able to meet Jühnde's mobility needs as well. And we would be in a situation where we could generate all energy free of CO_2. (Fangmeier, Jühnde)

Figure 3.3 Jühnde villagers in a meeting next to the biogas plant
Source: Bioenergy village Jühnde (2012).

The concept of bioenergy villages can be transferred to both developing and industrialized countries. However, it has to be locally adapted.

> Granted a certain transfer of know-how, bioenergy villages can also be realized in developing countries. They are always based on the concept of community, and one can use renewable resources in many places. However, the technology and the devices must be adapted to the specific features of the respective country. In the United States, for example, completely different technologies are in operation. Infrastructures also differ. Someone from Gambia is less interested in heating than in how to convert heat into cold or produce electricity from heat. (Fangmeier, Jühnde)

Information dissemination and Smart Community consulting (Jühnde)

The project in Jühnde has attracted a constant flow of visitors. The cooperative intends to use its achievements to establish a center that specializes in knowledge transfer.

> In the past, we had up to 8,000 visitors a year in our plant; today there are still around 3,000 visitors annually. We have hosted academics, politicians, agriculturalists, potential operators, housewives, etc., and we would like not only to offer favorable heating prices and to continue to optimize the project but also to pass on knowledge. Together with the society for rural adult education, the cooperative has founded a limited company called "Centre for New Energies," with the objective of professionally organizing the transfer of knowledge.
>
> We also pursue the idea to let firms exhibit their products and services in a dedicated information center, following a marketplace concept where they can get known to a wider audience. Some firms have already signaled interest, for example Claas, a huge manufacturer of agricultural machines. (Fangmeier, Jühnde)

The planned marketplace is not only targeted at regional and domestic interest groups and initiatives.

> About 20 percent of the visitor groups come from abroad. We already have good contacts outside Germany. Those contacts are highly valuable for firms, as they may initiate concrete projects. Opportunities in Germany are to a very great extent exploited and companies now see their markets internationally. Our academic partner, the University of Göttingen, is also internationally active. (Fangmeier, Jühnde)

Fangmeier is satisfied with the diffusion of ideas that was achieved by his community project.

> It was the right time to implement the first bioenergy village in Germany. Since then, the demand for bioenergy villages has been strong, with new concepts currently being developed. As a result of our village functioning as a role model, around 70 to 80 other villages have switched to bioenergy in Germany, and the number is growing continuously. Consequently, there are still some sales opportunities in Germany for companies, even though the greater potential for turnover may now be found abroad. (Fangmeier, Jühnde)

Similar to consulting services for municipal utilities on how to switch to a smart grid, Fangmeier observes that – beyond mere knowledge transfer – specialized consulting practices that give advice to interested communities have already emerged.

> A company from Southern Germany has already converted around 10 villages to bioenergy supply and is presently working on other projects. In that case, commercialization has worked very well. (Fangmeier, Jühnde)

However, the market has not yet reached a stage where Fangmeier would quit his stable job to launch his own company.

> In order to realize this business idea, I would have had to give up my job as a quality and project manager. Since the market was still uncertain, I would have had to take a huge risk, which I did not want to do. So I preferred to work on the bioenergy project in my spare time while keeping my regular job. (Fangmeier, Jühnde)

Industrialized countries may be able to learn from least-cost solutions in the developing world about how to establish island systems in the future. Michael Weinhold from Siemens expects intelligent solutions that may spread to the industrialized countries:

> We must adopt a global perspective. In countries such as India or in regions like Latin America, there is still a need for basic electrification. Electricity consumption per capita in India, for example, is a tenth of that in Germany. Electrification, however, can only be accomplished if a significant number of major power stations are built and the transmission grids are extended. Yet at the same time these countries are also establishing micro grids,

Box 3.2 **Masdar – island solutions under harsh environmental conditions**

Can island solutions with an autonomous, decentralized energy supply work in harsh climatic conditions? With 25 tons per year, the per-capita CO_2 emissions in the United Arab Emirates rank among the highest in the world – the average German emits less than 10 tons annually.

How can a society with a strong environmental footprint create a carbon-neutral way of living? The pilot city Masdar, 17 kilometers outside Abu Dhabi, pioneers urban structures that will shape cities of the future. It was initiated in 2006 as a subsidiary of the government-owned Mubadala Development Company and is intended to accommodate more than 30,000 inhabitants upon completion in 2025.

The city aims to be largely energy-autonomous. With sunshine all year around, solar energy production will become the backbone of the island system. The region's largest array of ground-based photovoltaic panels will be complemented by roof-mounted PV installations, while concentrating solar power and tube collectors will deliver the energy for cooling and warm-water heating. Waste-to-energy solutions for non-recyclable substances and geothermal energy are also under consideration.

Planning a new city from scratch allows for integrated solutions. The British architect Norman Foster has developed an urban cooling concept that draws on traditional housing styles of the region, including a narrow street pattern with angles and axes that optimize shading and huge wind cones that will provide natural ventilation for offices and larger buildings.

Figure 3.4 Proposed master plan of Masdar City
Source: Masdar (2012).

> The energy consumption patterns of the first inhabitants are closely monitored to increase efficiency. Showers, refrigerators, and lighting are switched on and off automatically, cooling is centrally controlled, and residential water use can be restricted. The city is car-free, but fully automated transport facilities are within 200 meters' walking distance from each home. Masdar serves as a laboratory of how human behavior can be adapted to sustainable resource use. Via its institutions, an institute inspired by the Massachusetts Institute of Technology and specialized on renewable energies, and IRENA, the International Renewable Energy Agency, the founders try to communicate their novel ideas and designs.
>
> But the financial crisis has also affected the master plan. Some ambitious projects have been scaled down or cancelled, including a hydrogen power project and large-scale use of roof-top photovoltaic panels. Meanwhile, desert sand storms decrease the efficiency of panels by up to 30 percent, and have to be cleaned manually. Despite numerous obstacles, Masdar is poised to provide a showcase and laboratory that envisions how urban life can reconcile economic growth with a low-carbon lifestyle.
>
> (*Sources*: Masdar City, 2011, Vidal, 2011)

which are frequently based on renewable energies. For example, diesel generators are linked to photovoltaic panels, wind turbines, and batteries for storage. These micro grids form small, autonomous biotopes. Not every load will be connected up to a network or a high-voltage grid because it simply does not pay to connect up every remote village by means of a high-voltage power line.

The micro grid structures in the developing countries are often part of a basic electrification effort, while the same technologies are also discussed in the industrialized world under the theme of energy self-sufficiency. We will be able to learn from the developing countries how this can be done in a cost-effective way. As those countries have limited financial resources, there will be very intelligent solutions, which we will probably be able to employ here. (Weinhold, Siemens)

Recommunalization

The concept of citizen value (SWK, SWU, GASAG)

While bioenergy villages often emerge due to the interests of individuals at the grassroots level in the respective communities, another group of players in the energy market has discovered the attractiveness of communal projects.

> Municipal utilities and regional suppliers have a high degree of customer loyalty. (Hose, ODR)

More than 800 municipal utilities operate in Germany, the so-called Stadtwerke. Apart from energy supply, their businesses typically involve community services like waste, water supply, and public transport. They own medium- and low-voltage grids. The larger ones have electricity and, sometimes, heat generation facilities in their asset portfolios. The geographic and cultural proximity to their customers allows for participatory processes on the local level, facilitates identification with initiatives, and creates positive marketing feedback for the utility.

In Germany the pattern of small-scale structures with 890 municipal utilities allows us to implement decentralized structures more rapidly. Municipal utilities can be pioneers of a new energy system. If decentralized systems prevail on a global basis, the structures in other countries will have to undergo a fundamental change. At present, however, it does not seem to me that the politicians are capable of the innovation required for existing structures to be completely overturned. Of course, it would be possible simply to emulate the decentralized systems developed in Germany, and to a certain extent they already exist in sub-Saharan Africa. But there we have only isolated supply systems and no completely networked electricity grid. That was the situation in Germany around 1880/90. At that time, electricity was required for the trams and street lighting to operate. There was no network linking different towns. The network will remain but we will once again move in the direction of isolated, decentralized systems. (Jänig, SWU)

Carsten Liedtke, speaker of the board of directors of municipal utility Stadtwerke Krefeld (SWK), explains his company's concept of "citizen value."

Decentralized energy production has always played a particular role for municipal utilities. Only it was ignored for decades because the business was operated locally by the municipal utilities, be it as remote heat delivery or CHP generation. Remote heating networks are often only very local and their production structures are often only decentralized. Large businesses have been involved only to a limited extent.

The issue of public participation is becoming increasingly important because through decentralized energies, the customer also becomes a producer. For municipal utilities it is an advantage that decentralized and regional structures are fashionable at the moment. For the most part these structures are in the hands of the municipal utilities and these again are mostly dominated by the city authorities. These smaller structures can create local jobs and enhance welfare.

We want to be perceived as a modern enterprise with the corresponding control and processing tools, but at the same time we emphasize regional values. We support local sports clubs and theaters. In our opinion, it is an intelligent use of our marketing budget. Through our regional focus we can judge in which associations we have to be active in order to come across as credible. By bonding with local politicians, the local business community,

> and the public, we distinguish ourselves clearly from other competitors in the market. We call this concept "citizen value." The major players do not have this advantage. They are not so networked or present on the spot. (Liedtke, SWK)

Traditionally, municipal utilities are involved in social activities and sponsoring on the local level. Projects in decentralized energy generation are a logical extension of their "citizen value" involvement.

> Together with the local savings bank, Volksbank, we carried out the Krefeld Solar Power project where we opened a savings account that was transformed into a corporate bond. Within a few days, we were able to raise €7 million. We have used that money to install around 3 MW of power in photovoltaic plants on roofs in Krefeld. The effect on our corporate image was very positive, while we were able to develop even closer ties with our customers. (Liedtke, SWK)

In the current German setting with a highly disaggregated market structure, coalitions and alliances between municipal utilities will gain in importance in the future.

> There will surely be some consolidation, but the future belongs to municipal utilities. Perhaps there will be fewer of them, perhaps they will be more closely interwoven, enter mutually beneficial cooperations, or launch joint projects with shared ownership structures more frequently. Mega-structures will be less likely than ties between individual municipal utilities from one region, for example in IT. Smaller regional platforms are certainly realistic and also make sense. We already have cooperations in the area of energy procurement and energy trading.
>
> When utilities have cross-holdings, part of their core influence has to be shared with others. But it will only rarely be the case that a complete municipal utility identity is surrendered. Since customers are turning again more strongly to their immediate surroundings, the small-scale image of a municipal partner makes sense. A municipal utility can be linked to big ones and benefit from their structures and processes, but it is important to preserve its regional identity. For example, we have acquired a stake in the utility of a nearby town and have taken over the management of commercial operations. (Liedtke, SWK)

The municipal utility Stadtwerke Unna (SWU) is another example of how municipal utilities aim to attract environmentally aware residents. Christian

Jänig describes the company's stance toward the environment and the constraints a municipal utility faces. Because of its multi-purpose character and public service obligations, it cannot become the cost-leader but has to differentiate its services in another way:

> In 1993 we raised ecology to the same status as economy among the objectives in our mission statement, and since 1995 we have issued an environmental statement. We regard climate protection and reduction of resource use as important, thinking about future generations. We have been advising our customers personally and locally for 20 years on how to save energy and can look back on 50,000 customer consultations. The savings made are up to 40 percent. To enhance customer loyalty, we have also been advising businesses for 10 years. This is important because SWU is quite small; we have to pay the treasurer an annual charge of around 40 percent and therefore cannot lead on price. We therefore try to keep our customers and have a low turnover rate of just 1 percent. We have managed not only to build customer relations but also to be recognized as an ecologically-oriented municipal undertaking.
>
> At the moment, we are still not able to store electricity in large amounts, since we have no pump storage plant. We rather distribute surplus energy cost-effectively to other areas with a virtual power plant. We use this system to optimize the total procurement of gas, heat, and electricity. Within three years it was already paid off. We can support the city of Unna with our profits, which means that the city can allocate funds for urban development. That boosts the prestige of our firm and anchors it even more firmly in the local population. Such things only work with a local basis. (Jänig, SWU)

The municipal utility of Unna also promotes carbon-free energy supply and climate protection projects.

> A decentralized energy concept or system is very beneficial for customer relations. We are being increasingly accepted since we have now been offering 100 percent renewable electricity for three years without any surcharge. We procure the energy at the exchange, which means we can optimize energy purchase and therefore sell it at favorable rates. For the town of Unna, we have developed a climate concept with 89 measures for the period 2010 to 2020, which we are now implementing. Apart from energy-related initiatives, it includes also measures for reforestation or open water lanes. (Jänig, SWU)

The proximity to the final consumers is the unique selling proposition that municipal utilities can offer. However, a gradual expansion into adjacent communities may be envisaged.

> We are already in the process of optimizing and coordinating decentralized generation plants throughout our concession area. However, we do not intend to expand further; we are simply too small. A basis of trust is important for us. In bigger cities outside our core territory, we are unknown and would not have a basis to get accepted there. Decentralized systems only function locally, analogous to the slogan "Think global, act local." (Jänig, SWU)

Andreas Prohl of gas utility GASAG agrees on the marketing appeal of localized energy:

> We want to create a closed energy world in which the customer is at the center and can learn and experience everything himself. In this way, we distinguish ourselves from those companies that can only provide abstract, intangible offers such as electricity from an offshore wind park in the North Sea. (Prohl, GASAG)

To create a whole energy world for their customers, GASAG is also considering integrating electric vehicles into their portfolio.

> In the future, we wish to become an all-round supplier, no longer just for gas. Owing to the high level of media attention, electro-mobility is very well-suited for conveying this message. (Prohl, GASAG)

Findings on empowerment and recommunalization

- *Citizen value is a key selling proposition for utilities*: Local utilities have the most favorable conditions to succeed in the decentralized market because they can offer an integrated energy world to their customers. Compared to supra-regional competitors, they are able to strategically differentiate their products and services due to their profound knowledge of the local context and the preference sets of their customers. They can enhance loyalty and consumer acceptance of new technologies by issuing local bonds.
- *Consumer empowerment needs agents of change*: The value proposition of bioenergy villages is their contribution to community life in a more holistic rather than narrowly energy-supply oriented way. The collective's motivation hinges on the idealism of interested individuals and volunteers, though, who are willing to invest a substantial amount of their time for the advancement of the venture.

Box 3.3 Dezhou and the Chinese Solar Valley – reconciling growth with environmental protection

How can fast industrial growth be reconciled with a sustainable lifestyle? After two decades of unprecedented economic performance, Chinese authorities have turned their attention to a better living environment for its citizens. Not only has China become the world's largest and most competitive producer of photovoltaic cells, the country is also improving its domestic record of energy use.

Dezhou is a city with more than five million inhabitants in northwestern China. It is the home of two major producers of solar thermal technology: Himin Group and Ecco Solar Group. In total, more than 100 solar enterprises have been established in the agglomeration, which led to coining the region the "Chinese Solar Valley." The products vary from simple flat-plate collectors to high-end vacuum-pipe collectors. Photovoltaic and energy-conserving glass industries are also located in the city. Greenpeace estimates that in 2007 around 800,000 people in Dezhou were employed in the solar industry.

The goods produced in Dezhou are not only exported, but also installed locally. More than 30 kilometers of the city's main streets have street lighting run by electricity collected during daytime with photovoltaic panels, in addition to the illumination provided for some public parks and squares. Many traffic lights are also powered by sunlight. The city has undertaken major efforts to promote solar devices with two programs: "One million solar roofs" and "Solar bathrooms in hundred villages." In 2010, 80 percent of urban buildings were using solar water heaters. More than 100 villages were equipped with solar bathrooms.

Figure 3.5 Solar thermal heating in Dezhou
Source: Su Li/Greenpeace (2010).

Instead of imitating products from other countries, Dezhou emphasizes that local scientific innovation is the key to raising the solar industry level of Dezhou. Until

> 2010, it obtained 586 solar product patents and undertook several solar scientific research programs. In solar thermal utilization, the share of the self-owned technology exceeds 95 percent. Dezhou's industry also innovates in related technologies like high-temperature electricity generation, seawater desalination, heating, air conditioning, refrigeration, and construction energy conservation. Dezhou solar industry is one of several national industries that completely possesses intellectual property rights in China. "This project is yet one more example of how China is moving rapidly to lead the global clean energy industry while the United States falls behind... China and other Asian nations will out-invest the United States in the clean energy sector by over three to one over the next five years," says Teryn Norris of the Breakthrough Institute.
>
> (*Sources*: Norris, 2010, Solar Cities Initiative World Congress, 2010, DZWWW, 2003)

- *Bioenergy villages serve multiple objectives*: In industrialized countries, the use of local biomass to provide electricity and heat is likely to remain more costly than the use of conventional primary energy resources in the near future. However, an increased autarky corresponds to most public authorities' long-term environmental objectives and policies. In addition, it creates new ties among the residents to their community and may contribute to reducing urban migration. Given that bioenergy will remain a niche market, sustained state support is likely to remain limited.
- *Smart community consulting goes global*: International interest in bioenergy villages is increasing; pioneers will serve as triggers and will communicate best practices. Local artisans, technicians, and SMEs can benefit from their experience and sell their expertise in both developing and industrialized countries. Dissemination networks share their experiences and engage in reciprocal learning.

4
Smart Management of Electricity and Information

Decentralized energy supply relies on the technical and logistical coordination of a multitude of microscopic market agents, often geographically dispersed and with fluctuating, rather unpredictable load and feed-in profiles. Especially in the grid-connected energy infrastructure, the instantaneous balancing of dynamic demand and supply patterns becomes a major challenge, and exceeds the complexity of previous optimization efforts by several dimensions. High-voltage transmission lines have to be reinforced to transport renewable energy from offshore wind farms to urban load centers, while the low- and medium-voltage grid has to bear additional electricity flows – either from photovoltaic panels installed in residential neighborhoods, or into the batteries of electric vehicles.

Intelligent metering and communication systems for final consumers have created opportunities to measure and manage demand response on a highly disaggregated level. Smart meters are being tested around the globe in pilot projects and have already experienced a massive rollout in some European countries. However, the initial enthusiasm has somewhat faded after the first field tests; utilities, policymakers and scholars now share a more differentiated view on the actual savings potentials. Where do business executives see future value creation in the intelligent grid and smart meters? Are smart grid and smart meters inevitable components of the future energy system, or just temporary hype that does not create lasting profitability beyond environmental marketing? How can utilities and market incumbents position themselves vis-à-vis new entrants from the information and communications sector? What is the role of regulation – should authorities impose a mass rollout of smart meters, or should this development be left to market forces?

Theoretical framing

Network externalities

Under current regulatory practices, in many countries, including Germany, investments into the transmission and distribution grid are subject to a restrictive rate-of-return policy. Network operators are not compensated for any additional investment beyond the minimum standards, although making the grid more intelligent appears an imminent necessity to manage decentralized energy feed-ins based on small-scale renewable energy supply. Similarly, smart meters may suffer from the apparent drawback that the investment and operation costs for prototypical residential consumers exceed the expected benefits. If the future grid were smart and every household had a smart meter – effectively participating in active demand-side management – would the overall welfare effects legitimize the additional investments and a mass rollout of smart meters? In other words, are there positive network externalities associated with a smart electricity system?

Network externalities have to be distinguished from economies of scale: The latter assumes that a heavy deployment of smart technologies will decrease costs for the equipment, the associated IT services, and maybe even for the installation due to the increasing expertise and competition among technicians. By contrast, positive network externalities may emerge from the fact that if more people have a certain device installed – say, for example, a specific home TV movie player – more products will be available in formats for that device. Each user of that device has private benefits, but also creates unintended benefits for existing users (for a more detailed discussion on the topic, see, for example, Lopatka and Page, 1999, Liebowitz and Margolis, 1994). Likewise, each local grid operator is able to monitor with greater transparency its area's demand and supply patterns, which is instrumental in increasing the stability of the overall network. Economies of scale could hence be considered a supply-side effect, whereas network externalities show characteristics of a demand-side effect. To what extent does that logic apply to smart meters and a smart grid?

While the network operator has to cope with costs for the smart grid that typically exceed its direct benefits, smart grid and smart meter investments generate informational advantages that primarily serve the coordination requirements of third parties. Market participants thus obtain additional benefits without bearing the full costs of installation and operation of the devices. Figure 4.1 by Müller et al. (2010) illustrates this discrepancy.

The following positive externalities can be assumed to occur with an intelligent grid, if final consumers receive adequate price signals (based on Mukherjee, 2008, So and Lagerling, 2009).

First, environmental conditions could be improved due to more coordinated electricity demand, in particular reducing peak loads; utilities may shift their

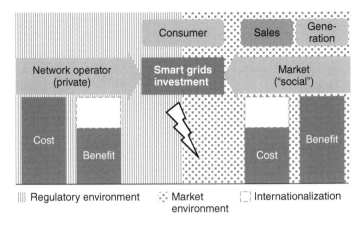

Figure 4.1 The externality problem with the design of smart grids
Source: Müller et al. (2010).

supply from polluting peak plants to times with abundant renewable feed-in; second, outages and power interruptions could be foreseen and prevented more easily, thereby reducing the opportunity costs for industry, commerce, and residential consumers.

In the longer term, infrastructure and grid investment costs would be lower than under a conventional grid replacement scheme because of enhanced transparency of the actual power flows.

Müller et al. (2010) suggest that the grid operator who invests in the smart grid also sets a network capacity price for local renewable energies producers to recapture additional costs. Those local producers benefit from less frequent technical feed-in restrictions. This mechanism would, of course, be effective only under a regulatory regime that does not oblige the grid operator to pay even for feed-ins that do not actually occur due to local network constraints.

The retailer would refinance its additional expenses for a smart meter by selling technical information of its customers to the network operator, who would benefit from greater transparency of the electricity flows in its demarcation area, for example with the objective of outage prevention.

However, a lower bound of minimum final electricity consumption of a customer may have to be defined. Interviews with industry experts, in particular smart meter producers and utilities, provide insights on where the critical threshold should be set from the practitioners' perspectives.

Standardization and lock-in effects

If an early standard is established, it may lead subsequent adopters to choose the same standard to ensure compatibility, and a positive network externality may emerge irrespective of whether the standard is the technologically optimal solution.

In the early stages of the evolution of a new technology, manufacturers compete for setting a *de facto* industry standard that ensures a sustained strategic advantage in the market. Office software is one of the examples where a dominant firm successfully applied this strategy. The network externality then creates increasing returns – and may ultimately lead to market failure. Once that a *de facto* standard has been established, the dominant firm can reap monopoly returns and recoup its initial expenditures.

Standards thus have the positive side effect of potentially reducing investment uncertainty for investors, but they may lead to lock-in effects in a suboptimal or unilaterally exploitable technical configuration. Is the smart meter industry prone to path-dependence effects created by a dominant firm?

Marketing research shows that within the global competitive landscape, indeed two companies, Swiss-based Landis & Gyr and US manufacturer Itron, sold more than 50 percent of all smart meters (smartmeters.com, 2011). A regional disaggregation, however, reveals that the European market is dominated by neither of these companies, but rather by California-based Echelon, which has market shares of around 95 percent in Italy, 60 percent in Denmark, and 40 percent in Sweden (Frost & Sullivan, 2011a). The company also promotes an application layer protocol that can be used with multiple communication media, the Open Smart Grid Protocol, which is run by 81 percent of European smart meters.

Echelon benefitted from a classic first-mover advantage because its technology was deployed in the by-then world's largest smart meter project, initiated in the first years of the new millennium by utility Enel in Italy and now serves almost 30 million homes. This early lead has given the company a competitive advantage in European markets, but not necessarily elsewhere.

Meanwhile, the North American market is dominated by three manufacturers – Sensus, Silver Spring Networks, and Elster – while in China fierce competition is reported to exist among almost 100 predominantly domestic companies. The country's largest producer, Jiangsu Linyang Electronics, held a share of less than 7 percent in 2010 (for a more detailed analysis, see e.g., Research In China, 2011). It is only a matter of time until electronics giants General Electric, Siemens, ABB, or Toshiba make serious attempts to enter their regional (and later expanding their global) smart meter businesses and either compete with the current incumbents or just simply acquire them, as Toshiba did with Landis & Gyr in July 2012. The entry of large industry conglomerates is likely to lead to industry consolidation.

While companies compete for market shares, in all major world regions diverse committees have been established to formulate standards for smart grid protocol inter-operability. In the United States, the Energy Independence and Security Act of 2007 assigned responsibility to the National Institute of Standards and Technology, while in Europe the European Commission and the European Free Trade Area issued Mandate M/441 to the European Committee for Standardization, the European Committee for Electrotechnical Standardization, and the European Telecommunications Standards Institute, which are in charge of forwarding suggestions. Under these multiple-agent negotiations, it seems likely that standardization of smart meters will evolve like cell phone protocols, where regional *de facto* industry standards compete with *de jure* standards. Given the diversity of the players, the standards are likely to be open, that is, not proprietary, and perhaps not even be patented.

The major challenge for the smart meter industry will be to ensure worldwide compatibility of their systems, leading to more complex technological features than under a single global standard, roughly comparable to tri-band or quad-band cell phones capable of using differing frequencies in North America and Europe.

Smart grid

The conventional electricity grid is characterized by a natural monopoly of the grid components, in economies of scale in generation, and through the political motivation to provide electricity as a centrally organized public infrastructure service. The instantaneous matching of supply and demand requires a sophisticated system architecture and coordination and optimization between the relevant generation units, which have traditionally been handled by vertically integrated utilities and their respective network operators.

The smart grid dissolves the unidirectional orientation of the grid – from the generating unit to the final consumer – and introduces a range of new control and optimization devices that assist in managing multiple, locally dispersed electricity sources and sinks, as well as stabilize frequency and voltage of the network.

Typically, the smart grid is considered the 21st-century response to a mere copper reinforcement of the existing lines. It includes wide-area monitoring and control devices like sensors and capacitors for distribution grid management, as well as communication equipment like routers, relays, switches, and IT servers. The objective is to achieve real-time transparency for substations and to create automation mechanisms to measure the activity of network flows.

In addition, superconductors for transmission enhancement, advanced metering, and electric vehicle charging infrastructure, as well as building automation systems, smart appliances, and in-home displays for customer-side systems and residential buildings are considered components of the smart grid (for a comprehensive overview of smart grid technologies, see IEA, 2011).

The major components of a prototypical smart grid are depicted in Figure 4.2.

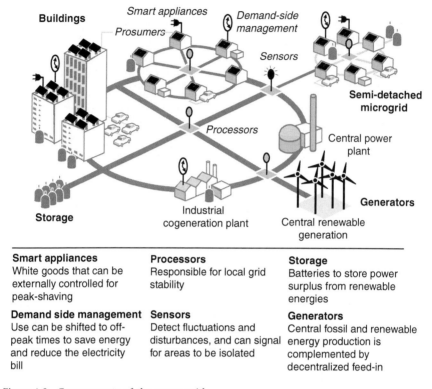

Smart appliances White goods that can be externally controlled for peak-shaving	**Processors** Responsible for local grid stability	**Storage** Batteries to store power surplus from renewable energies
Demand side management Use can be shifted to off-peak times to save energy and reduce the electricity bill	**Sensors** Detect fluctuations and disturbances, and can signal for areas to be isolated	**Generators** Central fossil and renewable energy production is complemented by decentralized feed-in

Figure 4.2 Components of the smart grid

However, the terminology remains vague, and it largely depends on the country context how political and corporate decision-makers define their smart grid. Michael Kirsch, in charge of intelligent grids at regional grid operator EnBW Regional, relates the differences in the definition to varying challenges across countries:

> The terms "intelligent grid" and "smart grid" are understood differently in different countries, since the challenges are also different. In the United States the highest electricity consumption is during the summer, but this is also the time with the least available generation capacity. Balancing out these two factors is their definition of an intelligent network. In Europe, generally speaking, the farther east, the poorer the network quality. The mere construction of load control centers is what those countries understand to be an intelligent network. The Italians have shifted the emphasis and installed intelligent electronic meters in order to compensate for the poor network and the country's severely inadequate generating capacity.

In most countries, the conversion of the conventional, copper-based electricity lines into a smart grid is motivated by two drivers – one being political, the other one rather technical:

First, an initiative by decision-makers to establish a network that utilizes state-of-the-art communication technology, with the ultimate objective to enhance efficiency, introduce smart appliances and demand-side management, and reduce carbon dioxide emissions; second, the technical necessity of grid operators to respond to challenges created by an increasing share of decentralized energy generation, which tend to destabilize the overall system.

Both dimensions interact on the regulatory level: If the government generously subsidizes the deployment of photovoltaic cells and wind turbines, the system experiences temporary reversals of load flows, local supply spikes, and more complex balancing tasks.

With government incentives and utilities tackling the challenge of balancing their networks, the market potential of the smart grid will unleash vast amounts of money. Consulting practice Pike Research estimates that the software and services that will enable smart grid data analytics will increase from approximately US-$350 million in 2010 to almost US-$4.2 billion in annual revenues by the year 2015 (Pike Research, 2010b).

The feed-in induced revolution (Siemens, EnBW Regional, and ODR)

One of the key components of the liberalization of European electricity markets has been the unbundling of generation, transmission, and distribution, and the introduction of competition in retail markets. In Germany, and many other industrialized countries every final consumer can freely choose the company that provides the power for his or her needs. The physical flows of electricity follow Kirchhoff's Laws, though. This implies that even if a final customer selects wind power from the Baltic coast as the source of her primary energy, the electricity that arrive in the household might have been generated in the nearby lignite plant. Nonetheless, the retailer has to pay fees to the respective grid operators, including a local network operator in charge of "the last mile". This utility is also in charge of coordinating the network flows.

> In Germany, we are faced with the challenge of installing large-scale wind parks far away from load centers, while placing far more than one million mini power plants on residential buildings stock. Bavarian barns and roofs comprise 20 percent of global photovoltaic output. In 2011, Germany's electricity capacity grew by another 7.5 GW using photovoltaic panels and 2.8 GW from wind turbines. We now have more than 50 GW of solar and wind capacity available, which has a privileged feed-in status but whose electricity production does not always fit actual demand. Before the emergence

of fluctuating renewables, it was possible to optimize easily the dispatch of big power stations, based on likely consumption, which could be fairly accurately calculated. But PV plants and wind energy are weather-dependent – we cannot accurately rely on them in winter, in cloudy weather, or when the wind ceases to blow. (Weinhold, Siemens)

With the emergence of these highly volatile renewable energies, the major challenge for utilities is to find new ways of balancing supply and demand, both over time and geographically.

We will shortly have a scenario in which we will have more renewable energy production in the network than domestic demand. Since we still have severe bottlenecks towards neighboring grids, there is an urgent need to expand the German transmission grid, in particular in order to be able to transport large quantities of energy from the north of the country to the load centers in the south. The German Energy Agency estimates that by 2020 around 3,500 km of additional transmission lines will be needed. Within recent years, in sum only 90 km have been constructed. The authorization and permitting procedures take around 10 years. We expect severe grid congestion. (Weinhold, Siemens)

The smart grid also includes innovations in the long-haul transmission grid.

A great deal has happened in recent years with regard to grid technology. For example, there are now direct current lines available that could transport wind energy over larger distances with minimal energy loss. At the same time, network control technology has advanced so that power outages can be almost totally avoided. In Germany we have the most secure power supply of all, with around 16 minutes of power cuts in the entire year. (Weinhold, Siemens)

The currently available storage capacities in the German system are not sufficient to absorb the fluctuating supply.

We have to increase flexibility on the load side to absorb the volatile quantities of energy. Since negative prices were permitted at the German electricity exchange two years ago, absurd incentives were being given to destroy energy. The German pump storage plants have a maximum capacity of only 40 GW – equivalent to two hours of strong wind. If volatilities increase, conventional power plants will also have to become more flexible. (Weinhold, Siemens)

For Michael Weinhold, large and conventional power plants will remain the backbone of German electricity supply:

> While small-scale investors foster decentralized, renewable energy supply, we continue to fundamentally benefit from the large power stations and the integrated network that we have. They ensure that sufficient energy is available. Conventional large power stations remain indispensable. It is possible to interrupt generation with fossil energy sources for a short time and to fall back on major storage capacities, but we cannot afford the complete outage of the power supply for a single minute because, in that case, many business enterprises would encounter massive problems. That is why every European country, despite the integrated European network, has devised its fleet of power stations so that the domestic power supply is, by and large, self-sufficient. (Weinhold, Siemens)

Weinhold sees business opportunities related to the smart grid, in particular in the intelligent operation of power plants and data management systems:

> Liberalization brought more players into the market. A politically intended disintegration of the stakeholder landscape can be observed. Generation and network departments operate independently of each other as a result of unbundling. Consequently, the strategies of stakeholders tend to increasingly diverge.
>
> Many industries are about to enter the electricity sector. For example, manufacturers that are specialized in control technology of conventional power plants now see their chance and are developing systems that allow for an extremely flexible mode of operation in order to compensate for the generation of volatile renewable energies. Other industries target the infrastructures of mega-data management, which are now established in the United States and other countries. (Weinhold, Siemens)

Manufacturing companies like Siemens may benefit from a competitive advantage because they can provide products and services along the whole smart grid value chain.

> Our Siemens portfolio encompasses mobility, automation, grid technology, and power station equipment and services. We can combine our competences and offer clients customized solutions. Our holistic view of the product range makes us a valuable partner for municipal administrations. Our services cover the whole energy system, from operating power plant for clients, servicing plants and networks, etc. There is an increasing number of stakeholders now who may own a plant but cannot service it.

> We are an enterprise that acts globally and we claim to be trendsetters. We are ready to cope with any company that tries to challenge our leading position. If we are not the leaders, we try to get into the top position in the relevant market segments. In Germany we have built the world's biggest and most efficient gas and steam turbine. The feeling of pride about our achievements lasts about a tenth of a second, though, and then we get back to work.
>
> We constantly take a look at the requirements in the various regions and, using our market intelligence, try to extrapolate trends from them. We conduct so-called pictures-of-the-future analyses, where we talk to many stakeholders and ask them how they think the future will look in 10 or 20 years. We then draw our conclusions and assess whether our innovation roadmaps are still accurate or whether we are on the wrong track. Since we operate globally, we can make comparisons between energy systems and learn from each other. We also look at adjacent industries – our market research is like a permanent radar screen. (Weinhold, Siemens)

Siemens has developed a new kind of innovation network based on open innovation:

> We promote the concept of open innovation. We organize so-called innovation jams where we ask external stakeholders to express their opinions and ideas on internet platforms. We would then like to integrate the core competences into our enterprise and develop them further. Today, innovation means collaborating closely in networks and teams. (Weinhold, Siemens)

Michael Kirsch from EnBW Regional, an operator of the low- and medium-distribution grid in southwestern Germany, describes the choice between reinforcing the old copper grid and moving to decentralized solutions:

> With respect to our low-voltage grid, renewable energies pose a huge challenge for us. For example, a village with 75 households and a population of 190 inhabitants has photovoltaic plants with a capacity of around 1.2 MW. We are now constructing a fourth substation there. When the grids were designed – only for consumption – we calculated that one substation was sufficient for 350 households.
>
> To cope with fluctuating energies, the distribution grid can either be reinforced with thick copper lines, which are extremely expensive, to transmit current over long distances, or begin to generate more electricity on a decentralized basis and make use of storage technologies. The grid operator would then be transformed into an energy manager who also provides backup

power. That would result in a change in the functions of the grid operator and a redefinition of the responsibilities.

We are in the process of transformation from an electricity supplier to an enterprise that absorbs all the renewable electricity that is produced locally. That is a strong driving force in the question of expanding and planning the grid. We must adjust to a greater extent to these changing requirements. (Kirsch, EnBW Regional)

Both transmission and distribution services are considered natural monopolies and hence regulated under a strict cost-based revenue scheme enforced by the German federal network agency, the Bundesnetzagentur. Its overarching objective is to ensure stable and secure grid operations, but it also acts as a watchdog to prevent local grid operators from exploiting their local market power.

The regulatory agency is pursuing two goals. It is aiming to ensure that the grid is free of discrimination and that competition is not obstructed. By and large, this goal has been achieved. The other objective is to monitor prices while ensuring that the energy supply is secure. The regulatory agency has massively intervened with regard to grid fees and licenses, but with modest success. Although grid fees dropped, electricity prices in fact increased. (Kirsch, EnBW Regional)

Kirsch comments that the incentive regulation imposed by the agency indeed generates efficiency increases.

We are very much governed by the conditions that the German grid agency imposes on us. As grid operators we have to observe precise conditions as to how much we can charge. Since we are obliged to install costly connections to renewable energy sources, we have a clear incentive to become more efficient.

We as grid operators have to take account of the political environment, regulation, the Federal Network Agency, and economic viability. While we wish to be innovative, we still have to develop economically sustainable concepts. In general, grid operators rather tend to be followers than first-movers. We at EnBW are perhaps innovative followers or slow first-movers. (Kirsch, EnBW Regional)

A regime too strictly imposed by the agency may negatively affect the transformation of the network into a smart grid, though.

The grid operators wish to take the lead in expanding a smart grid but need money to achieve this. Reinforcing the conventional grid would be much more expensive than a smart grid, since it would require huge quantities of copper. Generally speaking, both ways of extending the grid lead to an increase in electricity prices. However, this rise is ultimately lower if intelligent systems are used. The need for expanding the grid is now generally accepted and is promoted by the German government and the EU. We would very much welcome if at least the costs incurred in establishing intelligent grids were recognized. (Kirsch, EnBW Regional)

The positive network externalities of a smart grid may hence not materialize because of a mono-dimensional focus on preventing so-called X-inefficiencies, the misallocation of public resources in regulated industries (see Leibenstein, 1966, for a further discussion of the topic).

Local grid operators face the trade-off between profitability and the obligation to connect all local producers to the grid.

Plants with power exceeding 100 kW$_{peak}$ are regulated by the feed-in legislation to ensure the system stability. Smaller plants have to be negotiated by individual contracts to avoid voltage problems in the low-voltage grid. Local grid operators are caught between an obligation to connect and considerations of economic efficiency. We have to expand the grid and try to accelerate the process in the most intelligent manner. (Kirsch, EnBW Regional)

However, new legislation allows for controlling and disconnecting larger loads.

Legislation envisages that the new energy law will specify that there will be a reduced grid fee for loads that can be switched off, such as electric vehicles. There is likely to be a provision in the Renewable Energies Act (EEG) that plants larger than 100 KW$_{peak}$ of installed capacity can be controlled by the network operator and automatically taken from the grid if necessary. This actually implies phasing in an intelligent grid. (Kirsch, EnBW Regional)

The gradual evolution from a regional service provider to an active agent in the energy transformation is exemplified by regional distribution company EnBW Ostwürttemberg DonauRies AG, located in a rural area in the southwestern part of Germany.

Electricity used to travel on a one-way street, from a few big power stations along the extra-high-voltage grid to the high-voltage grid, the transformer station via the medium-voltage grid into the low-voltage grid. Grid control was primarily necessary at the high-voltage levels. Since the low- and medium-voltage grids were subject to little control, we have only very limited data information on them.

But now we have thousands of decentralized installations. Photovoltaic panels feed current into the low-voltage grid, while wind power and biomass feed into the medium-voltage grid. If there is too little wind or not enough sunshine, then the electricity flows in the classic way to the consumers, namely from the big power stations, which are hence still needed. But if the sun shines and the wind blows, many households become producers. This is a task that the grids must be designed for. We have to forecast the amounts of energy and know which quantities will enter the grid; when and at what connection point; where the power will flow; and how it will be marketed. The smart grid will deliver this information. The grid infrastructure will have a primary energy grid and a communications network as a layer on top. Communications and energy systems will become increasingly intertwined to obtain information on the load and feed-in profiles of every decentralized supplier, and grid-specific data such as voltage levels. Transformer stations, which previously had no communications link, must be controlled.

In big cities such as Berlin and Stuttgart, the grid is stable and strong because there are many industrial customers and multi-family buildings with a high population density and correspondingly high energy consumption. There are many transformers and substations, stable power lines, and few feed-in points such as CHP units and photovoltaics. But in rural regions with agricultural areas and a dispersed housing structure, you will find single-family houses, barns, and extensive plots of land, which means a lot of space and plenty of sun but a low load. If three to five times the load is fed into the grid in those places, the existing network will be totally inadequate. The medium-voltage level must be brought closer to these feed-in points. Substations and existing power lines have to be reinforced and new power lines and additional transformer stations built. Voltage used to be regulated in the transformer stations; now that has to be done in individual substations. The challenge for grid technology is to configure the grid so that peak power supply can be borne. (Hose, ODR)

In Germany, more than half of the installed capacity of renewable energies belongs to private owners and farmers, while local utilities and large energy companies have a share of less than 20 percent. Figure 4.3 on the following page shows the evolution of renewable capacity according to different owner groups.

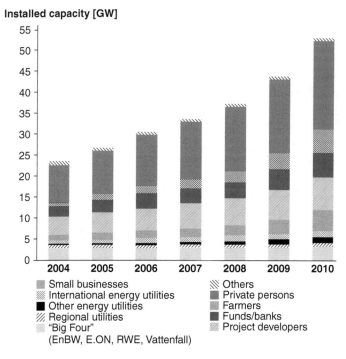

Figure 4.3 Ownership of decentralized energy generation units in Germany
Source: trend:research (2011).

Decentralization has gained lots of momentum and cannot be stopped. It is not the energy utilities that mainly triggered this transformation but private persons with photovoltaic installations on their roofs or with a stake in wind parks, or farmers with biogas plants.

In our service area, more than 90 percent of photovoltaic plants are in the hands of homeowners and farmers. However, EnBW has also initiated and realized many projects where locals have a stake in photovoltaic plants on public buildings, such as schools. (Hose, ODR)

Energy utilities have realized that a parallel, decentralized supply structure has evolved and will continue to grow, as long as the government intends to support them.

The feed-in tariff under the Renewable Energies Act has triggered investments of around €20 billion in 15 GW of photovoltaic plants within the last two years. They contribute around 4 percent of electrical energy in Germany on an annual base, but already more than 20 percent on sunny days; 30 to 40 percent of the photovoltaic panels have been installed in Bavaria. The average Bavarian now has 500 watts of power from photovoltaic modules, resulting in 400 kWh of electricity production on average per year. A household of three to four persons consumes around 3,500 kWh a year. So Bavarians meet at least a third of their electricity in private households, with the average of 500 watts distributed from photovoltaic modules. The nationwide average in Germany is around 200 watts. (Weinhold, Siemens)

In the distant future, renewable energy installations will be cost-competitive compared to centralized power plants due to cost degression in manufacturing of photovoltaic panels.

Of the approximately 220,000 ODR customers, there are already 21,000 who generate electricity with photovoltaic installations on the roof. In 2020, we predict that a quarter of our customers will have their own production facilities and another quarter will have financial stakes in local photovoltaic or wind power initiatives. Since the feed-in remuneration is being reduced, there will be increasing numbers of small producers who will be catering for their own consumption.

In the ODR we have an electricity demand of around 3,000 GWh. In our region more than 30 percent of that is produced from renewable electricity – in balance sheet terms: produced from private photovoltaic plants and from biogas and wind power installations. Around the year 2020, this figure will even rise to 85 percent on some days.

While we have expanded hydroelectric power, photovoltaic, biogas, and wind power, we have done so only to a relatively small extent. We need wind power installations because they offer the last chance for any kind of participation in renewable generation technologies. Through the phasing out of nuclear energy, we lose capital and yet still have to invest in the smart grid, smart meters, and the expansion of wind power and in storage capacities. For too long, the big players concentrated on centralized systems. They underestimated the dramatic increase in the number of producers of renewables. (Hose, ODR)

Hose expresses concerns about future power traffic jams on the transmission highways across Germany.

Major investors like the big energy utilities erect large-scale wind parks in the North Sea and the Baltic Sea. But the main load centers are located in the south or west of Germany. We have to transmit all the electricity via transmission grids that do not yet exist. (Hose, ODR)

The business model of a local utility like ODR may be endangered if decentralized energy generation continues to expand.

> If a household has a mini CHP plus a photovoltaic plant plus a storage unit, it has a completely autonomous energy supply and probably does not need an electricity connection. If the system progresses too quickly, we will not be able to sell a single kilowatt hour. Then we would practically be a grid operator that only has to transmit very small amounts of electricity within its grid to provide balancing services or to supply customers who are not self-producers.
>
> The grid will become uneconomical and grid fees will rise. As a result, this will inspire even more customers to become their own suppliers. A central grid will only be economically viable if all consumers pay a basic charge, like a service fee corresponding to the actual grid fee. We are lucky that the output of the installations of private producers is often far higher than the electricity they consume themselves, even if they are able to store parts of it. That is why they have to transmit some of it via the grid. (Hose, ODR)

Hose describes the challenges his utility faces:

> EnBW, ODR's mother company, originally focused on hydro-electric power among the renewable energies, especially run-off river plants and biomass power stations. Then wind power emerged – initially in our home territory in southern Germany but later with onshore wind turbines in northern Germany. Now wind parks are being built offshore in the Baltic Sea. The first of them, Baltic 1 with 50 megawatts, is up and running. Around 40 municipal utilities were involved, including the EnBW Ostwürttemberg DonauRies AG, ODR for short, which is my employer. Now more wind parks in deeper waters are being planned in the Baltic and later also in the North Sea.
>
> The state government of Baden-Württemberg aims to have 38 percent renewable energies by the year 2020 and 86 percent in 2050. This involves a major program for expanding onshore wind power in the region. Some years ago, EnBW set up a dedicated renewable energies unit called "Triple E," which plans and builds these wind parks and also photovoltaic plants. In the ODR area, there are some excellent wind power locations where up to 300 wind turbines can be erected. Together with "Triple E," we will build more wind power plants in our region. We are currently in discussions with the population, the local authorities, and the landowners about finding sites not affected by protests and under consensus between ordinary citizens, local mayors, and district authorities. As a supplier of decentralized

solutions, those endeavors should be profitable, but one must not focus only on financial returns.

In order to promote wind power, not only in our region but also in other places such as the Black Forest, which also has areas that are very suitable for wind power, we plan to set up a company in which municipal utilities, the municipalities, and ordinary citizens participate. The EnBW, or rather "Triple E", is taking over the project planning while the regional centers or the ODR, for example, maintain regional and local contacts. They know the people on the spot, district administrators, state and regional planning authorities, mayors, local citizens, and also local investors, above all the private banks and publicly owned savings banks. In our region alone, nearly €1 billion could be invested in this way. That is beyond the capability of my company. In addition, the wind parks need substations and power lines to transport the energy that is generated. It must be explained to people how such projects can be integrated and how they affect the region.

The local population has to bear the aesthetic burden and put up with the other inconveniences caused by wind power installations. Consequently, locals do not want to supply distant cities but retain the electricity in the region and economize on grid expansion. The savings banks, for example, set up a fund so that local citizens and councils can also identify with the big wind turbines. (Hose, ODR)

The future configuration of the grid will not only concern and affect grid operators, but also owners of renewable energy installations. Hose expects that local power markets will evolve wherein individual customers will sell their electricity:

In Germany the load fluctuates between a 30 GW minimum and a maximum of 90 GW. There are now around 55 GW of available power from renewable energies. Periods when there is an abundance of electricity will become more frequent in future.

At present, the grid operator absorbs the electricity from the producer and passes it on to the transmission grid operators, who collect it and market it at the electricity exchange. The deficit is balanced by an assessment of all electricity consumers.

In the future, electricity will be traded and marketed on the decentralized level. On average, in a rural area like ours, the size of PV installations is around 25 kW, producing around 25,000 kilowatt hours of electricity annually. If it is a single-family house, that building actually requires only 5,000 kWh for itself. If it is supplemented by energy efficiency measures, it needs perhaps only 3,000 kWh. So what will it do with the other 20,000 or 22,000 kWh, if in future the owner only receives remuneration for part of it or perhaps no remuneration at all? The owner will have to market his electricity directly.

A municipal or regional utility, which has installed the metering facility, could offer to purchase the excess energy. At the electricity exchange, the price is around five or six cents, or tomorrow perhaps four cents, and I can buy up the current for a little less.... Then I have benefitted by a margin, while the local producer also has a profit. That would be decentralized, regional trade beyond the system of big power stations and the marketing of huge amounts of electricity. (Hose, ODR)

Alexander Voigt, founder of island solution developer Younicos, views the future grid as loosely connected subsystems that only occasionally exchange larger amounts of electricity:

The grid of the future will consist of many zones that, in themselves, can be insular and that have a percentage of generation within the zone and will buy in energy from neighboring zones for certain times and months. Naturally, even systems with a high degree of self-sufficiency will in future still be linked to an overarching network. (Voigt, Younicos)

Christian Jänig of the municipal utility Unna also predicts that a move towards a more decentralized structure will take place:

Consumption and generation must be coordinated by intelligent systems like virtual power plants. Yet politicians are not yet fully aware of this. It will take a few more years until changes are brought about in Brussels because other countries are still lagging behind. That hampers our efforts to protect the climate. Even a rise in temperature of just two degrees could have a massive impact. So we must act. Decentralized energy systems offer an opportunity. We will certainly have to retain a few large power stations for voltage and frequency stability, but certainly not the same amount as today. (Jänig, SWU)

Regulation should provide sufficient incentives for investments in a smart grid:

Investments in the smart grid are not recognized by the Federal Network Agency. The company is therefore not reimbursed. Nonetheless, we are currently running pilot projects to gather experience with the smart grid. The fundamental issue is optimizing grid stability. Ideally, processes should function fully automatically in the control center. For example, we test whether and how we can regulate or even switch off renewable energies. (Jänig, SWU)

> **Box 4.1   A smart grid for a Smart City – Boulder paves the way and shows the obstacles**
>
> The university town of Boulder, Colorado, will become the United States' first "SmartGridCity." The pilot project, which has received international attention, was launched in 2008 by local utility Xcel Energy, together with partners like Accenture and management software provider GridPoint. The project foresaw to equip residential, commercial, and light industrial customers with smart meters on a voluntary basis. Out of more than 45,000 connections, 43 percent of Boulder residents had smart meters in 2010.
>
> Residential customers are able to use their smart meters to view their electricity usage in 15-minute increments via an online web portal for better daily management and control over their electricity use. They can also create a personal energy profile that instructs the home's appliances and devices, such as thermostats and pool pumps, to automatically manage individual loads according to their customized preferences, like seasonal, work or vacation schedules. All homes obtain a back-up battery storage unit that buffers power for almost two days in case of an outage. Electric vehicle owners are able to use a smart charging device to automatically charge their cars during off-peak hours. In addition, photovoltaic panels are installed on rooftops. Excess renewable energy can be sold to the utility.
>
> For Xcel Energy, the pilot project entails, of course, the possibility of tracking every household's consumption and perform meter readings automatically, but it can also provide insights about which energy-management and conservation tools customers want and prefer, and which technologies are most effective in improving the way they deliver power. However, rolling out the planned smart-grid components on a wider scale faced some severe challenges. Most importantly, cost overruns amounted to three times what the utility had planned to spend. Together with its industry partners, Xcel Energy invested more than US-$100 million. One of the reasons was an explosion of the costs for the fiber optic network, which is used for SmartGridCity's information transfers. Far more underground fiber was needed than anticipated, and the local geology required specialized machinery. Fehrenbacher (2010) criticizes that a proper cost-benefit analysis, which might have revealed some of the complexities, was not undertaken before the beginning of the project. In addition, the metering systems already installed would not provide as many benefits as anticipated.
>
> "Xcel relied on fiber-optic cable, which is notoriously expensive to install, rather than on newer, wireless technologies, which have improved dramatically since the Boulder project was started. Then, too, Xcel focused first on the big infrastructure pieces of the project, which are required to upgrade the aging grid itself, rather than on consumers, who had hoped to see early benefits in their own usage and pricing," Time magazine journalist Amy Feldman comments.
>
> Even if the project turns out to have been too ambitious and costly, it positively contributes to the learning curve with the smart grid, and may serve as an example for other municipalities about which mistakes one can avoid when trying to set up a smart grid.
>
> (*Sources*: Feldman, 2011, Xcel Energy, 2012, Fehrenbacher, 2010)

ICT to manage bidirectional power flows (Itron, Argentus, and ODR)

If the smart grid is established, the local utility may turn into an energy manager. As industry analyst Marianne Hedin of consulting practice Pike Research rephrases the challenge: "The 'data tsunami' that will wash over utilities in the coming years is a formidable IT challenge, but it is also a huge opportunity to move beyond simple meter-to-cash functions and into more robust business intelligence capabilities, true situational awareness with real-time optimization of their operations, and even predictive analytics." She expects smart grid utilities to evolve into "brokers of information" (Pike Research, 2010b).

Implementing the smart grid is a major challenge, especially for smaller utilities. Many of them have been overwhelmed by the success of the feed-in regulations that resulted in a boom in decentralized energy generation. The corporate structures had to be adapted accordingly. Michael Kirsch describes the challenge a regional grid operator like EnBW Regional faces:

> We separate our business units according to the voltage level. The extra-high-voltage network with 220 or 380 kV is run centrally by the official grid operator. It is predominantly concerned with transmission issues and has major customers such as municipal enterprises. It also coordinates various pump storage plants. By contrast, the low- and medium-voltage grid is managed by decentralized sub-units. They were classically geared to customer consumption and grid customer service. Due to the expansion of renewable energies, the issue of decentralized feed-ins has gained huge importance. Since the law on renewable energies has come into force, we have actually been overwhelmed by the quantities of decentralized energy that has to be borne. We had to build up our manpower and work hard on the process structures and the IT, which has to operate without frictions and discontinuities. (Kirsch, EnBW Regional)

The smart grid requires municipal and regional utilities to implement sophisticated methods to manage fluctuating, dispersed supply patterns. One effective means to accelerate the organizational learning process is to establish field tests.

> We run a pilot project with smart meters. We installed them in a number of households. They continuously measure voltage and consumption levels. This data is then transmitted to the local grid station via a power line and from there via GSM or DSL to us. We then have the possibility to control and switch devices on and off. In Freiamt, a village with 5,000 inhabitants, we have established our so-called grid laboratory. Freiamt's electricity is supplied via a transformer station, which helps us to accurately estimate which quantities of electricity are consumed and how much is locally generated. We test various applications in the medium-voltage area, such as the coordination of renewable energies production, communication, and IT. We also use the pilot project to refine our estimates of weather forecasts and photovoltaic intake in the low-voltage grid. We have various smaller-scale

projects focusing on smart control technology, and the storage of photovoltaic current in batteries. Our next project is a smart substation with an adjustable transformer. (Kirsch, EnBW Regional)

The configuration of the smart grid is closely linked to local demand and production patterns. EnBW Regional's strategy is to establish a so-called toolbox:

> For our toolbox we founded the "Grid Concepts" department within the EnBW Regional AG because local requirements can vary greatly: In one location, a big wind farm may have to be connected to the high-voltage network, while in another location individual wind turbines have to get connected. In fact, they are best linked to the medium-voltage network. We will also have to deal with challenges arising from the future topic of electric vehicles. (Kirsch, EnBW Regional)

EnBW Regional also offers its expertise to local initiatives and associations.

> All our activities have to be reconciled with our core function as grid operator. Through the unbundling directives, we are not allowed to sell electricity or photovoltaic plants or mini CHP units. That is the responsibility of our parent company's sales department. However, with our know-how we can support local authorities and local energy cooperatives with their projects. (Kirsch, EnBW Regional)

Eckhardt Rümmler from E.ON expects three types of service providers to emerge:

> In addition to billing, new players from data processing have all the data at their disposal and could offer specially tailored, individual products for the respective customers. For the financing of distributed plants, a third player could enter. Ultimately, there will be technology-driven, data-driven, and customer-product-focus-driven players in the market. (Rümmler, E.ON)

Instead of only selling the new technological devices, a new business opportunity emerges in advising and assisting grid operators in the transition, according to Karsten Peterson and Werner Paech from smart meter provider Itron:

> Regulation foresees that smart meters have to be installed in new buildings and as part of major renovation work. This leads to the situation that even small energy supply companies, municipal utilities, and suppliers that may have only a mere 10,000 customers are immediately compelled to employ a technology that does not yet correspond to what the standards will be in the future.

There is nothing to prevent a municipal utility from deciding today to undertake a complete rollout with the technology now available. After all, meters with communication technology will be usable for the next 10 to 15 years, even though a different protocol may get established. But the task of selecting a smart meter from among a range of very different suppliers is a challenge for smaller municipal utilities. There are very many different technologies on the market; for example, the communication can be implemented via the electricity network, mobile communication, or DSL broadband. (Petersen and Paech, Itron)

Petersen and Paech observe an enormous consulting potential in the fragmented German utilities' market segment:

In Germany we have about 800 energy suppliers of widely different sizes. At least 500 of these 800 enterprises are faced every day with the question of what meters and what measuring technology they are to use when meters have to be replaced on a routine basis because their validity has expired. Will it still be possible to use the Ferraris meters next year? And if not, which smart meters would be suitable? Not every municipal works has in-house IT experts who are familiar with communication technologies, power line carriers, GSM, mobile phone technology, and with the relevant tariffs of the providers.

While my company has not yet started selling these consultancy services at the moment, we have the great advantage that with our meter business, we have been in close contact with our customers for years. We are informed about the internal structures of our customers and make use of our experience. In contrast to many of our competitors, who only take on the job of installation and initiate the data transfer, we know where the data can be fed into the system. We are in a position to supply the logistics and not just the meter. Supplying meters will continue to be one of our core businesses, but in the case of municipal and medium-sized energy suppliers, we can create new business lines of handling a whole range of complementary services. For these enterprises it often makes no sense to carry out the task themselves, especially not at present where everything is in a state of flux. Recommunalization also implies that ever-smaller suppliers are established, and the business orientation of utilities is changing.

To read the meter data and to administer and control the meters, a certain minimum size, minimum security, and also performance of the IT employed is required – irrespective of whether 50 or 5,000 meters are in the network. Apart from supplying the meter, we therefore offer customers the logistics around the meter as a service – at least for a transitional period, in order for the customer to become familiar with the issue of smart metering and to make meaningful use of the meters. (Petersen and Paech, Itron)

Konrad Jerusalem, CEO of energy performance contracting provider Argentus and co-founder of Kofler Energies, anticipates potential consulting services also for industrial consumers:

> In the coming years, measuring electricity flows will become a major component of energy costs. The trend will go beyond small smart meters and move toward more sophisticated devices. (Jerusalem, Argentus)

Instead of outsourcing, ICT services may compensate municipal utilities for the declining returns created by competition and grid regulation:

> Commodity sales and the grid fees will provide us with our minimum returns, but the topics for tomorrow are services connected with energy efficiency, storage, and energy control systems. We might to some degree cannibalize our own core business of selling electricity if we speed things up. But if we do not step in now, we will be out of the game tomorrow, and others will do the job.
>
> Losses through customer consumption and energy efficiency must be offset by control services and through consultancy services, including for industrial companies. They hand over the control and metering system and data protection to those they know, who are well-established and reputable. That is our opportunity at EnBW. In data management, we can offer the collection of the gas, water, and heat meter-data via a single system as a service to the municipalities who do the water supply and the municipal utilities who provide the gas supply. If regional utilities lose shares in the electricity market, it is a way to secure new value creation. (Hose, ODR)

Competition for utilities will increase, both from telecom companies and municipal utilities:

> In the electricity market, the classic competitors will remain, although competition will be rendered less attractive by the self-producers. We will continue to see strong competition between municipal utilities. Those municipal utilities in big or medium-sized cities have the huge advantage that they supply a very compact area and do not have the problem of expanding the grid because they have relatively strong power lines and industrial customers and they are integrated. They enjoy an incredibly high degree of customer loyalty. They also expand into neighboring areas, form mergers, and open up their own utilities, whereas rural suppliers are stuck with highly demanding rural areas.
>
> Other competitors will include the telecommunications companies. They can handle the issue of billing with various tariffs and flexibilities better than energy supply companies. If they become cheaper than us in metering operations and data management, including gas and water, we may lose

the metering business to them. Since commodity sales are no longer very attractive and we can no longer pick the cherries out of the grid, we may face a serious problem.

When working together with municipalities and producers, one gets a better idea of the current developments. The municipalities have already realized that if they hold on to their grid, they will at least have secured a minimum return. Competition for us comes from the efforts by municipal utilities aimed at re-municipalization. (Hose, ODR)

Regulatory unbundling and the division of ownership along the energy value chain, in particular the separation of distribution services from the retail contact with the customer, has received some criticism from the network operators. Synergies related to an intelligent coordination of all decentralized agents may actually be lost:

> Unbundling in that case might be counterproductive. If a sales division operates independently of the grid, this is counterproductive. The owner of a photovoltaic installation with a storage unit must be incentivized to feed his midday peak into the storage unit so that there can be savings in grid expansion. But if the sales department does not care about the grid and just wants to sell electricity to the customer, it has to adapt its strategy to the customer's wishes and allow him to consume electricity whenever he needs it. If the sales department is not able or allowed to develop a joint product with the grid operator, then we are lost. Grid and sales must sit together at the same table, in line with compliance. Gas supply, electricity supply, the communications systems, IT, grid – all parties must be involved. That is one advantage of municipal utilities or regional suppliers like ODR. The customer is completely indifferent as to whether we are unbundled as long as he derives an advantage from it. And for him there is only the EnBW or the ODR or the enterprise as a whole.
>
> The best example is the development of smart meters. The smart meter was assigned to sales because it was supposed to be a sales and marketing product to save energy. The grid was excluded from considerations first. But in reality a neutral grid meter is needed: first, for the billing of all customers and to substitute smart grid solutions; second, it will become a gateway for the customer to provide smart home solutions. (Hose, ODR)

Smart meters

Competing standards (Itron)

While the smart grid encompasses a whole range of new technologies introduced to balance and optimize load flows over the whole network, smart meters exclusively target final customers and are a pre-condition for active demand-side management. In the public discussion, however, smart meters have attracted far more attention than the smart grid because they directly affect the habits and routines of residential consumers – via an increased awareness of individual consumption patterns, data protection issues related to the outsourcing of the control of shiftable loads to private firms, and by the lack of profitability of a smart meter investment for many smaller consumers.

Like the move from regular cell phones to smart phones with a more ample functionality, smart meters represent the shift from the traditional electricity meter to a multi-functional device.

Petersen and Paech from smart meter producer Itron sketch the historical development:

> Smart meters entered the public debate around the year 2006. At that time, the objective was to cover more functions with the meters than previously possible. The classic Ferraris meter has a proven track record of functionality and quality. But if energy consumption is to be broken down according to the time of the day or even individual appliances, the capabilities of the traditional meter do not suffice. In the first stage, the classic mechanical metering technology was replaced with equivalent technology on an electronic basis. First-generation electronic meters had no greater variety of functions than mechanical meters.
>
> However, at the time, the functional scope of the new meters had not yet been precisely defined. They were meant to enable the customers to control and check their energy consumption instantaneously and to have an interface that would make remote meter reading possible once the IT infrastructure was set in place. (Petersen and Paech, Itron)

Regulation should reinforce variable tariffs and allow for differentiated pricing incentives:

> A future smart meter or measuring system must inform the electricity supplier or retailer about consumption patterns of the end user. It should have some type of remote functionality that enables the supplier to adjust and control the meter, while it should also be able to transmit a price signal to the end consumer, such that the consumer knows when electricity is expensive or cheap. In addition, household appliances should be able to communicate with each other, obviously according to the final customer's preferences. The refrigerator should know when the oven is switched on, and then automatically turn itself off. Strict tariff schemes should be abandoned, and enterprises should be permitted to introduce flexible tariffs and flexible price zones. On the bills for final consumers, the kilowatt hours consumed, along with their respective prices, should be clearly indicated. (Petersen and Paech, Itron)

As a consequence of the lack of globally agreed protocols, differing technologies, and the multitude of players, standardization of the new devices is only in its initial stage. Regarding the hardware, two solutions for smart meters are currently in the discussion, and it is not yet clear which solution will become the dominant design.

> There are two differing directions of smart meter development. The first one separates the communication from the meter – a concept called "meter plus MUC," where MUC is the acronym for Multi Utility Controller, now called Gateway, as a reference to its functionality. It is a separate communication unit and can collect and encode measurements via standardized and secure interfaces from all types of metering devices, including gas and water meters. A remote interface transmits the data to the energy supplier, the retailer, the grid operator, or any other entity that is in charge. It also facilitates the in-house communication, that is, the display of the data on the home computer, the TV, and mobile devices. The other option is to combine the meter, communications modules, and the transmission technology into one single compact device. (Petersen and Paech, Itron)

According to Itron's smart meter experts, both options have advantages and downsides.

> The integrated version has the drawback that the energy supplier and the grid operator are not necessarily part of the same company. In addition, there may be a third party involved who is solely responsible for the metering technology and the data collection. With this fairly complex constellation, it may be difficult to collect data from all meters with one integrated device.
>
> If the metering technology and the communication devices are separate, one major disadvantage is that additional interfaces are required. The data connection from the meter to the MUC/Gateway must be specially secured with regard to data transmission.
>
> Two completely different appliances with their own network devices and interfaces have a substantial price disadvantage compared to an integrated device, in which components and construction groups can be used together.
>
> However, both approaches are already in use. (Petersen and Paech, Itron)

Figure 4.4 shows a potential solution to combine gas, electricity and water meters, and the IT and communication devices, as well as the linkage to a back office, where the data are monitored and managed.

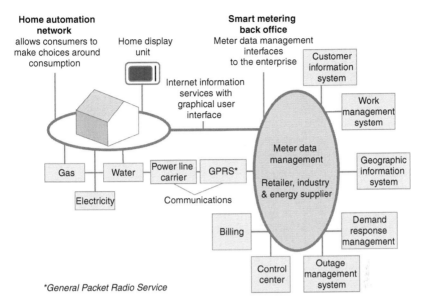

Figure 4.4 Smart metering solution by Siemens
Source: adapted from Siemens (2012).

In addition to technological diversity, smart meter data interfaces and communication protocols are not unified. Until a uniform standard is established, the test and pilot phase will continue, and current solutions are likely to be transitory. Protocols can even differ among countries that are closely linked by supranational regulation, for example within the EU.

In Germany, a data protocol has been established, which is known as Smart Metering Language (SML). Germany actively promotes that protocol in international negotiations. All appliances to be introduced to the market in Germany in the future will function according to this standard. In Europe, other countries push for harmonization based on a different standard called Device Language Messaging Specification (DLMS). The open questions of protocols and interfaces and the implementation of data security in these interfaces and protocols show that we are still in a test- and pilot phase. Current solutions will be transitory. Every smart meter installed today will not meet the standard that will prevail when all data protection requirements are taken into account. (Petersen and Paech, Itron)

An early mass rollout may lead to path-dependencies under a suboptimal design.

> Since the requirements of smart meters will change, it is questionable whether it makes sense to impose the compulsory installation of such devices before a stable technical configuration has been reached. (Petersen and Paech, Itron)

The representatives from Itron argue against a too laid-back position of the government. They comment that smart meters will not be voluntarily installed unless financial benefits for the final consumer are apparent:

> Over the last few years, the regulator was hesitant in taking decisive action. Legislation does not impose a mass rollout but assumes that the market will trigger the diffusion of smart meters. However, it has not been recognized that a market for a product only develops if the customer who has to pay for the product also derives an additional benefit from it. In the case of smart meters, the extra benefit for the final customer often occurs after smart meter deployment, so the benefits are slightly delayed. (Petersen and Paech, Itron)

Too restrictive regulation may lead to a rollout of sub-optimal technologies:

> Meters are becoming commodities with steadily falling prices.
>
> The greatest obstacle for a voluntary rollout by the utilities is that legislation sets a limit of not more than €10 per year to be passed on to the final consumer. The result of this cap is that there is a demand for less expensive products from abroad. If a grid operator does not have the option of charging the customer a higher fee for a meter with more sophisticated capabilities, then – as long as there is no mature business model – the network company will opt for a cheaper, low-quality meter. The law must give those who are obliged to install meters a chance that the investment will pay off. Otherwise, the energy utilities will not support the rollout of smart meters on their own behalf. (Petersen and Paech, Itron)

Standardization and coordination among market agents do not lead to an unfavorable and premature lock-in, but help to reduce uncertainty and drive down capital expenditures, according to Michael Kirsch of regional grid operator EnBW Regional.

> Standardization brings cost benefits and security for the industry. We seek to discuss and resolve as many issues as possible within the respective business associations to find joint solutions and to develop common standards. Within our industry association, we have actually achieved an agreement on standards for smart meters, which we as grid operators would like to install. (Kirsch, EnBW Regional)

Our interviewees from smart meter manufacturer Itron favor an incremental approach for the rollout:

> Lawmakers should stipulate an obligatory load profiling for industrial and commercial customers. This is already compulsory for large industrial companies. Paradoxically, the threshold of minimum annual consumption was raised from 30,000 kWh to 100,000 kWh a few years ago. Fortunately, this regulation was reversed again. A new German law ratified in 2011 requires measuring systems for those customers with consumption of 6,000 kWh or above. An average two-person family in an urban apartment consuming around 1,300 kWh a year clearly drops out of the target group.
>
> Similar to mobile phones, customers will switch appliances on or off at certain times when different tariffs are introduced. But a smart metering system is most useful above consumption of 10,000 or 15,000 kWh a year, for example for bakers or final customers who have demand spikes that can be shifted. (Petersen and Paech, Itron)

Regional grid operator EnBW Regional also questions whether a government-imposed mass rollout of smart meters is reasonable. For the moment, an implementation in rural areas with renewable energy intake seems more necessary than in urban areas, unless other systemic features like electric vehicles are added:

> It can be doubted whether we need a mass rollout of smart meters, even if logistically it would be the simplest solution and would also have certain cost advantages arising from economies of scale. A proper cost-benefit analysis could give us some indications.
>
> Contrary to the assumption of the Federal Network Agency, the market itself does not regulate where smart meters are introduced.
>
> In villages where every second roof has photovoltaic panels, a smart meter that both measures and transmits the voltage and also controls the feed-in of the photovoltaic plant is a sensible investment. By contrast, in cities with many large apartment blocks and single households, a smart meter rollout is not yet necessary. However, things can change if, for example, electric vehicles hit the mass market. Then consumer load control will become a serious issue even in the cities and make a rollout necessary in areas of high population density. (Kirsch, EnBW Regional)

The growth in the smart meter market is predominantly created by government policy. According to consulting practice Frost & Sullivan, Europe will experience annual growth of 26 percent for smart meters over the next years. For example, the French energy regulator has decreed a mandatory smart meter rollout by 2016. The British government intends to install 53 million electric and gas smart meters in homes and businesses by 2019 (Frost & Sullivan, 2011b). McKinsey predicts a tripling of smart meters in the United States by 2014, with a total of 50 million smart meters installed (Booth et al., 2010). In China, the State Grid Corporation plans to construct a power information collection system in its 27 provincial branches by 2015, including the installation of 500 million smart meters (Research In China, 2011).

Some countries, like Italy and Sweden, have already implemented a mass rollout of smart meters, while other countries are currently in the testing phase or have set time targets for the rollout, like the United Kingdom and France. The German government, by contrast, remains highly skeptical vis-à-vis a mass rollout, doubting the overall benefits of the scheme.

Smart meter deployment follows differing rationales across countries. In contrast to Germany, where direct advantages for residential end-users may be limited, Italy's nationwide switch to a smart metering system has yielded substantial benefits.

> If the aim is to supply intelligent electricity products with variable tariffs, so that customers can orientate themselves to availability in their use, then smart meters are urgently necessary. The energy industry should enter the new era that we already have in communications. It should not be imposed by regulatory policy. If a private residential consumer wishes to have 100 percent renewable energy on more favorable terms than previously, then he will perhaps – all by himself – get the idea of installing a smart meter. Or he might have a committed landlord who makes the investment.
>
> Nevertheless, at present there are relatively few commercially meaningful applications for smart meters in Germany – very much in contrast to Italy, for example. Italy already has over 30 million smart meters; and many people there do not pay their electricity bills. With a smart meter, the energy supplier does not need to send a person to cut off the supply for these households, but can manage this operation centrally. Once the money has been paid in, the lights are switched on again. The Italian example demonstrates that smart meters can already have a major economic effect today. (Voigt, Younicos)

Apparently, one of the major hurdles in the potential rollout of smart meters is the consumer. Unless it becomes clear that a smart meter actually yields benefits, consumers hesitate with the installation:

> Most residential consumers are not yet ready to get acquainted with the issue of smart meters. As long as it is not clear to them that a smart meter brings benefits, they will be reluctant to accept the extra costs involved. At present, smart meters have no financial advantages for them, even if they

substantially change their consumption patterns. Above all, smart meters today increase transparency with regard to daily energy consumption. Customers become aware of their typical consumption patterns. However, it is not yet realistic to expect them to change their daily habits – yet here smart meters may contribute positively.

It would already mean some progress if consumers realized how much a kilowatt hour actually costs them. Similarly, most consumers have no concrete idea of the energy value of a kilowatt hour. If this changed, the next step could follow. It seems that the politicians and the energy industry have taken the second step ahead of the first and will have to lower their expectations. (Liedtke, SWK)

In addition, reading the smart meter displays may prove difficult for some consumers:

> Because the collecting and billing process become more and more complicated and the customers accidentally mix up the readout of the three metering-registers, it will probably create confusion. So the reading must be done automatically via a smart meter. (Hose, ODR)

Field tests have verified that the possibility of shifting and altering residential electricity demand is fairly limited. Petersen and Paech criticize that expectations of the savings potentials of smart meters were exaggerated:

> The initial communications with the final customer made the mistake of constantly stressing the potential offered by a smart meter for saving energy in households. This potential for saving energy does in fact exist, but at present it seems exaggerated. Only in exceptional cases, savings of 25 percent can be achieved by using smart meters. Lights are switched on when it is dark – and not when a lot of electricity is available because it is windy or the sun is shining. The oven in the kitchen is used when one wants to cook at midday or in the evening, but not at midnight because the electricity is cheaper then. There may be a certain potential for economizing, but it is unlikely that smart meters alone will lead to a substantial conservation of energy – for a cross-section of the population it is 5 percent at the most, rising to a maximum of 10 percent in some cases. (Petersen and Paech, Itron)

The initial savings potential may hover at around 5 percent, confirms Michael Kirsch, of EnBW Regional. In the longer term, though, the consumers' willingness to pay attention to electricity consumption habits even decreases:

> We have investigated to what extent customers can be flexible in their electricity demand. For example, price traffic lights show the customer how

much electricity costs in each moment. The amount of energy that can be shifted with these traffic lights amounts to about 5 percent of the total. Unfortunately, half of this figure vanishes due to habituation effects. In the United States, daily divergences in the prices per kilowatt hour are wider and create stronger incentives to shift quantities of energy according to the time of day. (Kirsch, EnBW Regional)

Pilot projects did not fully yield the expected benefits.

Our sales department engages in a proactive approach to customers. Their experiments with smart meters yielded mixed results. Perhaps we were ahead of our time and thus had to learn the hard way. However, our company does not intend to surrender. (Kirsch, EnBW Regional)

Ralph Kampwirth of electricity retailer LichtBlick explains that his company did not enter the smart meter market because it did not see the positive effect for the consumer.

Micro CHP units and "swarm electricity" are a genuine smart grid solution. The smart grid is an extremely trendy topic, but one can see that many concepts are still at the project phase and need time to mature. Smart metering is far from being economical for the customer. That is why we do not offer it. It is hard to estimate how quickly intelligent energy supply can be realized. That will depend above all on the overall political environment and on how costs develop. (Kampwirth, LichtBlick)

Eckhardt Rümmler explains why energy incumbent E.ON has located "smart and distributed" in its sales strategy:

In the sales area, for billing systems it is becoming ever faster and simpler to choose the most favorable tariff available. Smart systems will be especially popular with customers who see it as trendy to work and play with energy in the home. We have therefore taken the decision to integrate the topics of "smart" and "distributed" into our sales strategies. (Rümmler, E.ON)

Konrad Jerusalem, specialist in energy efficiency and contracting services, sees a main psychological obstacle in consumer data protection and privacy issues:

An ideal electricity network would enable the grid operator to switch individual consumers off when necessary. This requires access to the customers' private installations. It can be doubted whether end customers are willing to reveal how much electricity they use and at what times. That information could be used to draw conclusions about their lifestyles. The data protection angle is still not resolved. (Jerusalem, Argentus)

Frank Hose of ODR comments that, for example, municipal utilities, either by themselves or in cooperation, enjoy credibility and trust due to their local embeddedness:

> With a smart meter, consumption patterns could be recorded every minute. Smart meters create total transparency. That is why a reputable company is needed that takes data protection seriously. In that respect, the municipal utilities and regional suppliers, and also EnBW, have a strong position compared with a company a customer is not familiar with. (Hose, ODR)

Netting peak-shaving and increasing the share of flexible power demand (Itron, SWU, and Siemens)

Smart meters constitute an important component in a decentralized energy system. They enhance consumer awareness while increasing transparency of network flows, including information on load and feed-in patterns that utilities need to manage the local grid.

The positive network externality discussed in the introductory section of this chapter may have direct repercussions on the financial performance of the municipal utility. In a competitive retail market, every final customer is allowed to freely choose the electricity provider. Switching rates in the electricity sector are not as high as in telecommunications, and local utilities benefit from "consumer stickiness." In 2009, the German federal network agency recorded only 5 percent of residential electricity consumers proactively switching their provider, but switching rates of lucrative commercial and industrial customers rose to almost 16 percent in the same year (Bundesnetzagentur, 2011). This creates a high degree of uncertainty in the provision of adequate amounts of electricity, even in a medium-term planning horizon. In particular, if a utility *under*estimates the amount of electricity needed to supply its customer base, it has to undertake ad-hoc purchases on the electricity wholesale market. In peak times, these purchases may turn out to be more costly than a rollout of smart meters for peak-shaving.

> Smart meters may be valuable for an energy utility if electricity demand exceeds the quantities purchased via long-term contracts, and the utility has to buy electricity on the spot market. Then the smart metering device can substantially reduce costs, although not via the final customer but by smoothing out demand. (Petersen and Paech, Itron)

When smart meters are used for system control and optimization by the respective grid operator, they justify their investment costs:

> A smart meter with communication infrastructure makes the most sense when data are transmitted to and are meaningfully used by the grid operator. (Petersen and Paech, Itron)

The real benefits of smart meters become apparent only when combined with the smart grid because they create a positive network externality:

> A German energy provider has marketed a device that can transmit consumption data every 15 minutes or even in shorter periods. The final customers interested in such a device are a fairly small, technology-savvy clientele. It can certainly not be assumed that a system, simply because of its mere existence, will create great interest and be used on a wide scale. A final customer who pays €79 for a device and another €25 a month or €50 per year on top of the electricity bill has no chance of achieving savings of that magnitude. How can such appliances establish themselves in the market when at present they offer nobody an advantage because of the lack of infrastructure? Smart meters can only be seen in the context of the strategy of how we would like to build up the energy supply in the future. A smart meter is a precondition to phase out nuclear power, to decentralize energy, to use renewable energies, and to prevent a mere duplication of the existing copper grid.
>
> A large part of the population would understand that exact knowledge about consumption and production patterns on the household level helps to make better use of existing grids. The grid operators need smart meters to know where energy is being fed in. Regional network companies have to frequently halt wind parks when the wind blows strongly. (Petersen and Paech, Itron)

One major feature of smart meters is the possibility to observe energy demand of final consumers in great detail, especially if they turn into decentralized producers. In the conventional electricity system, local operators have based their estimates of aggregate household consumption on preconfigured load profiles:

> As grid operators, we have a precise knowledge of the aggregate load flows – it is not necessary for us to know the demand pattern of an individual household. We rely on standard load profiles that are adapted to regional, temporal, seasonal, and weather conditions. (Liedtke, SWK)

With the emergence of this type of "prosumer," grid operators require more fine-grained information to manage their network. Pilot projects help them to better understand the consumption and supply patterns of end users in the new decentralized energy system:

> Up to now we have been pretty ignorant of the consumption patterns of our residential customers. We were of course able to detect peak loads, but we did not necessarily need to specify who caused them. In a field test, we have installed smart meters in the homes of staff members in order to access certain appliances in their households. We can define, for example, when the freezer is switched off and switched on again. Once this system works without faults, we can start the mass rollout. We will deploy a new software, the so-called intelligent agent, which exerts control over household appliances fully automatically and reliably. These sophisticated control systems become necessary because our consumers are becoming prosumers. Compared to the past, they now assume three different roles: via micro CHP units they generate electricity, they may have electrical storage units in their houses, and they still consume electric power. (Jänig, SWU)

Since residential energy demand has been proven to be fairly inelastic in the current setting, irrespective of increased awareness of final consumers due to smart meters, utilities may have to explore opportunities how to make demand more flexible.

> In order to reduce discrepancies between renewable generation and demand, consumer behavior must be influenced. For private customers we will operate energy conservation schemes and thus somewhat intervene in the energy management of the household. The benefit for the private customer is, apart from energy and cost savings, a reduction in the complexity of the coordination of appliances. (Jänig, SWU)

Frank Hose of regional utility ODR sees advantages in controlling electricity flows more easily with smart meters:

> The major competitive advantage of ODR or of EnBW is the extensiveness of the energy data. Whereas previously only the standard load profiles of the household customers were available, we now need the feed-in profiles of wind power, biomass, photovoltaics, and a fully automated metering system. This consists of at least three metering devices: a feed-in meter that is directly attached to the photovoltaic installations and measures electricity generation; a consumption meter-register; and a meter-register that monitors the difference in flows into the grid.

It is a great challenge for IT and billing systems because then at least three meter-registers will have to be read and a typical prosumer will have to receive adequate bills that take a whole range of rapidly changing promotion and remuneration schemes into account.

We automatically obtain all the data with a smart meter. This enables us to monitor but also to exert control and switch profiles and tariffs. We can even switch on or off the storage units, heat pumps, and electric heating.

Our customers can download their load data from us. It is interesting for them to observe their auto-production with their photovoltaic panels.

The objective of EnBW and ODR is to strategically occupy the metering service and not to relinquish it to anybody else because then we know the feed-in and consumption patterns of our customers. Even if an outside trader were to come and try to sell electricity, it would not be an attractive proposition for him because he does not know the load profile of the customer or what he feeds in.

Economically speaking, it does not make sense for a trader to install its own meter when the metering business gets liberalized and standardized, even if some enterprises are actually doing it. If the customer – for example due to energy efficiency measures – consumes less, and the margin has shrunk anyway due to competition, then the local craftsman who replaces the meter wipes out two years of margins. In two years' time the customer may have already opted out of the contract. (Hose, ODR)

The incremental rollout of smart meters creates new business opportunities for meter producers. The German subsidiary of the US company Itron has a long domestic tradition in manufacturing meters but it is refocusing now on IT services and consulting.

We were originally a part of AEG and were then taken over by Schlumberger Group. After meter technology was divested from Schlumberger Group, we produced meters under the name Actaris. At the same time that the upheaval in meter technology began in Germany, we were taken over by American company Itron. Coming from the American market, where networks there are less reliable than in Germany and where a substantial market for smart meters has existed for quite some time, we greatly benefitted from the experience and competence of Itron.

Our company profile changes from mechanics and production of meters toward electronics, software, and communications technology. We turn into a provider of IT solutions. Of all company sections, only our IT department has been expanded and will continue to expand further. Our IT experts are in charge of the interfaces between meters, communication devices, and networks. They implement relevant standards together with our central development department. At the moment, we have around seven IT engineers especially for Germany.

Whereas classic Ferraris meters are read once per year, smart meters generate terabytes of data. Itron has developed and employed the necessary software in America to handle this data; interconnections with SAP systems and others are therefore possible. We are introducing these solutions to the German market.

A completely new line of business has been launched by our data center. It redistributes the data retrieved from the meters to authorized persons and companies. The data can also be fed into the billing systems of enterprises. This increases the attractiveness of smart meters for enterprises below a certain size. In addition, customers get remote access to the center and can monitor their consumption via their home PCs.

Smart metering provides a viable business case to other areas such as gas and water, and there is just one common contact person for several interfaces. One of Itron's subsidiaries manufactures gas meters in southern Germany. We also supply water meters. We can therefore supply the whole system for measuring network-based supply infrastructure from a single source. (Petersen and Paech, Itron)

Petersen and Paech consider a high qualification of their company's employees as being essential for their success in the market:

To expand our consultation services, we send our staff to trainings and joint courses in the United States, England, and France. Of course, not all staff members have to be competent in communications technology and IT. In our sales department, we employ only staff with university diplomas in engineering. Those sales managers will continue to offer our classic range of products to major companies. (Petersen and Paech, Itron)

Close cooperation between the sales department and the business development department is needed to provide tailored solutions for each customer.

Our salesmen are familiar with the structures of the energy utilities. They closely monitor the situation with regard to smart metering. On the basis of this information, our business development department produces a tailored analysis of what concept and strategy would be most appropriate for each utility. Then the sales department and the business development department present this solution. (Petersen and Paech, Itron)

As Michael Weinhold of Siemens suggests, one viable option would be to strengthen the role of heat pumps in the domestic heating system:

> In the long term, it is no solution to take renewable energies from the grid because of transmission bottlenecks. Smart grid technologies help to shift loads and make them more flexible. However, private households are fairly inflexible. The bulk of the load in German households is caused by consumer electronics. One can hardly tell a consumer not to watch television until the middle of the night because otherwise the grid is overloaded. It is important to make intelligent use of heating in electricity because electrical energy can be leveraged via heat pump systems. It will probably not be possible to boost the number of heat pumps quickly because that goes hand in hand with more efficient building insulation. (Weinhold, Siemens)

After a first peak in the first half of the 1980s, the sale of heat pumps in Germany caught up again after 2006. Under adequate government incentives, the association of heat pump producers predicts sales reaching 300,000 heat pumps per year in 2030. Under a less favorable regulatory framework, the current figures of around 50,000 pumps would still gradually rise to above 100,000 pumps per year after 2020 (BWP, 2011).

The association of the electricity industry in Europe, Eurelectric, suggests in its "10 Steps towards smart grids with a look at flexible loads" that switching off heat pumps could be controlled centrally in return for heat pump users receiving a preferential tariff (Lorenz and Mandatova, 2011).

Eckhardt Rümmler of E.ON also considers heat pumps a potentially interesting field for peak-shaving.

> We believe that demand-side management will play a major role. From the refrigerator to the heat pump we see substantial volumes that can be realized via peak-shaving. In combination with renewable generation, they offer the opportunity to balance out the system internally. (Rümmler, E.ON)

Peak-shaving and demand-side management may have a negative impact on the profitability of central power plants.

> Through the use of renewables, there will be fewer and fewer operating hours for the big conventional power stations. Such power stations used to operate for 6,000 hours, but they now have 500 or 1,000 hours fewer, and in future another 1,000 hours less. This renders them less economical and must be offset by a regulatory intervention with a component that covers the fixed costs of these power stations. Otherwise, there will be no incentive to build a power station. (Hose, ODR)

Smart home and cross-selling opportunities (Siemens, E.ON)

Even if smart meters may not become profitable as stand-alone devices for many residential customers, they may be part of an integrated strategy to deliver a "smart home."

Electronics companies like Toshiba or energy company RWE, in cooperation with software producer Microsoft, already design system automation for residential buildings, promising a combination of heating management and control of household appliances. For example, the approach of RWE and Microsoft SmartHome contains less than a dozen intelligent devices, like thermostats, sensors, and remotely controlled plugs, which residents can install in their homes without any previous technical knowledge. Any household appliance and the overall heating system are connected with a central controller via a wireless network, which optimizes the energy flows.

Figure 4.5 shows the vision of a smart home by US company General Electric, including a charging station for an electric vehicle.

Cross-selling opportunities also occur with communication service and entertainment companies to use the existing infrastructure and enter the market. For the representative of Siemens, green homes and assisted living will drive the use of smart meters, directly and indirectly:

> All sorts of companies and industries have nowadays access to the final customer. Communication providers enter a typical household through telephone or UMTS contracts. Companies like game providers reach

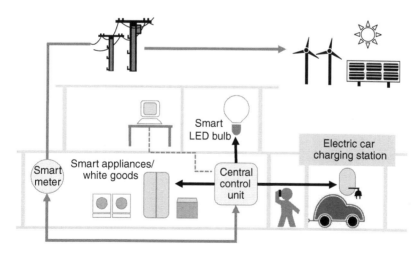

Figure 4.5 Smart home vision by General Electric
Source: General Electric (2012).

their very powerful consoles via the internet. Adjacent industries clash with each other and look for the business case in households. But smart metering is tricky: How can a telecommunications provider slash the electricity bill of the end customer if there are not many loads that can be shifted?

Soft factors matter much more. For example, how ecological is the lifestyle? The environmental footprint is important, especially for the younger generation.

By contrast, companies like Google count more on the convenience aspect. Google now offers a service via a cloud application, where a little chip is integrated into household appliances. With these chips, it is possible to switch on and off loads in the home with a mobile phone application. Perhaps the smart home ought to be renamed the "green and convenient home."

For an aging population, so-called assisted living will become important. Many people wish to stay in their own homes for as long as possible. In a few years, we may have an electronic assistant that will be our companion and support. Opportunities for cross-selling could emerge.

In general, new competitors do not necessarily appear in the energy system as we know it, in generation or grid services, but they also try to enter via the load side. Like parasites, they can rely on existing infrastructures and do not need to develop or install any hardware, but may still be able to create additional value for the customer. (Weinhold, Siemens)

E.ON demonstrates integrated solutions in smart homes.

> Which individual technologies play a part here is developed in some of our 12 technology and innovation centers. Half of them are concerned with small and distributed solutions, such as generation, storage, and smart applications in the household. We build demonstration houses in which smart homes are thoroughly tested. Other areas concerned are control devices, batteries, electro-mobility, and smart grids. (Rümmler, E.ON)

A holistic view of the modernization of buildings is more beneficial than the sale of sophisticated but overly specialized instruments that only tackle a fraction of the overall renovation needs, according to Johannes Hengstenberg of web-based building efficiency platform co2online.

> Metering service providers do not recognize the potential of their business opportunities. They see their strength in selling increasingly expensive metering devices and in some extra sales with complicated and more accurate meter reading. Instead, they should realize that they have the

data to approach the house owner with a plan for comprehensive modernization. A similar situation occurs with smart metering. Meters can be read with accuracy down to splits of a second and fitted with remote control, but the human interface has not been adequately taken care of. (Hengstenberg, co2online)

Municipal utilities see new business opportunities in integrated energy management.

> The first hurdle lies in the mindset. We must get used to having regional supply systems again instead of centralized ones. In order to avoid discrepancies between renewable generation and local consumption, we must influence consumer behavior. The next hurdle is regulation. EU-wide agreements will take some time. Until then we must tread a narrow line and provide the sales department with information from the network operator. The sales department can then offer contracts that, for example, guarantee that the customer will get by on €500 in heating costs per year. We will develop additional products and services. We will offer integrated energy management for private households within the framework of capped-costs contracts. Even if we slightly intervene in the household, we decrease the complexity for the private consumer. (Jänig, SWU)

The micro CHP units that LichtBlick sells as part of the swarm electricity strategy are part of a larger, integrated concept of establishing an energy world for the final customer:

> We think that in the future there will be far more integrated and smarter solutions for energy supply. There will be many other products and services based on intelligent energy generation. We are already considering how to expand our concept of "swarm electricity." We envision the future with a micro CHP unit in the basement, an electric car parked outside, photovoltaic panels on the roof, an energy storage facility at home, and the whole package integrated into the electricity market. It remains to be seen what is realistic. (Kampwirth, LichtBlick)

Findings on smart management of electricity and information

- *Renewable energy subsidies force the smart grid to be built*: If governments promote the deployment of renewable energy technologies, they inevitably reinforce decentralization. Operators of distribution grids already face major challenges to integrate and coordinate locally dispersed generation inputs.

> *Box 4.2* Telegestore in Italy – benefits of a mass rollout of smart meters
>
> While in many countries governments decide against a mass rollout of smart meters, because the savings potentials of residential customers would not justify the expenses for the devices and the related information and communications technology, Italian energy utility Enel voluntarily started in 2001 with the installation of 33 million smart meters. The remainder of regional distribution companies followed after a regulatory decree in 2006. By 2011, practically all Italian households had switched to a smart meter. Enel spent more than €2 billion on the installations. However, according to the company the investment has paid off relatively quickly. Enel estimates that annual savings due to the smart meters reach €500 million. The major reason for the financial success is the possibility for reducing non-technical losses – colloquially speaking, energy theft – and to automatically disconnect non-paying customers after a two-week warning period, in which they receive just enough electricity for basic appliances. The operating expenses per customer have thus decreased from €80 in 2001 to €49 in 2008.
>
> The smart meters are also used to monitor grid performance. System outages more than halved over the same time span from 128 minutes per year to 56 minutes per year.
>
> Enel introduced differentiated pricing schemes from which customers could choose. For example, the scheme "Sera" offers a 16 percent lower electricity tariff for the evening hours between 7 p.m. and 1 a.m. The scheme "Weekend+" has a 22 percent lower rate for Saturdays and Sundays. Enel estimates that 57 percent of all customers changed their behavior. For example, 29.3 percent of the survey respondents stated that they moved the usage of white goods into the evening, and 7.5 percent of those sampled switched off electronic appliances instead of leaving them in stand-by mode.
>
> When assessing the success of the scheme, it has to be taken into account that Italy faces a specific situation, though, because its non-technical losses may not be similar to those in other industrialized countries. In developing countries and emerging economies, where energy theft is not an uncommon phenomenon, the instantaneous, bidirectional information transmission capabilities of a system with integrated smart meters may prove to be attractive, because a leapfrogging effect may occur – a country could potentially omit one classic development phase in the electricity system and jump directly from no electrification to a smart meter, without the introduction of Ferraris meters.
>
> (*Sources*: e-Business Watch, 2009, ICER, 2012, Cotti, 2008)

An analysis of the costs and benefits of establishing a smart grid, as opposed to reinforcing the existing grid, often leads to decisions in favor of using smart grid technologies to tackle fluctuating demand and supply patterns. In order to accommodate fluctuating supply, regulation should focus on incentives to foster the use of appliances that allow for greater elasticity and flexibility in residential demand, which is less costly than reinforcing the existing grid.

- *Local electricity markets will lead to sunk costs in the transmission grid*: The expansion of main axes of the long-range transmission network is necessary to connect offshore wind power farms with distant load centers. However, once decentralized, largely autonomous island systems emerge, the amount of energy that has to be transported via some parts of the grid will substantially decrease. Investments in the reinforcement of the existing transmission infrastructure or the construction of new transmission lines should be scrutinized in a cost–benefit analysis that takes island systems explicitly into account. They could be avoided if proper incentives for decentralized supply were implemented.

- *Unbundling hinders seamless services*: The obligatory separation between grid operators and retailers leads to a suboptimal configuration of overall network flows because of differing incentives and business strategies. While the retailer has an interest to sell a maximum amount of energy irrespective of the timing of the delivery, the grid operator has to cope with peak network flows without being able to give proper price signals to the consumer or prosumer. If re-bundling is politically not desired, sophisticated regulatory instruments have to be established to enforce the optimization of network flows, increase consumer price elasticity, and ultimately avoid rate increases due to unnecessary investments in the local grid infrastructure.

- *Smart meters are instrumental to create local electricity markets*: For most residential consumers, the costs of a smart meter exceed the expected benefits. Load-shifting potentials are too small for an annual demand of less than 10,000 kWh. However, utilities and grid operators can benefit from information beyond prototypical load profiles to optimize network flows and apply peak-shaving and demand-side management. Automated reading and cut-off capabilities can substantially reduce operating expenses, especially in market segments where non-technical losses are high. These positive externalities have to be taken into account when assessing the full potential of smart meters.

- *Integrated consulting and IT services for municipal utilities will become a new source of value creation*: Current smart meter technologies are likely to be only transitory solutions because a dominant technology with unified protocols has not yet been established. The current diversity of available systems, as well as sophisticated communication technology, leaves smaller utilities in a state of uncertainty about which system to choose and how to use the information provided by smart meters. Specialized consulting companies can enter this new market and offer initial advice as well as monitoring and data management services to smaller utilities, whereas larger utilities can build up own expertise and sell it to peers. Local distribution companies may compensate for eroding market shares by offering value propositions before new entrants from ICT industries occupy that field.

- *Smart homes are the access gate for new entrants in retail*: Smart homes allow diversifiers to obtain easy access to the final customer. Telecommunication, entertainment and internet companies enter the market with their sector-specific expertise and uncover multiple cross-selling opportunities for bundled services. Manufacturers and specialized retailers like Siemens or LichtBlick will offer products ranging from environmentally sound demand-side management tools to services for special requirements like assisted living and charging services for electric vehicles.

5
Local Storage Solutions

Decentralized energy systems will at least partially consist of on island systems with interconnections to the central grid but with a high degree of autonomy. In these constrained settings with a high share of fluctuating energy inputs from renewable sources, matching supply with demand will become more challenging than in a larger technical configuration, where a portfolio effect is more likely to balance renewable energy intake and backup capacities based on, say, gas-fired power plants are at disposal. If largely autonomous patches of grid infrastructure are established, adequate investments for facilities that store excess electricity have to take place.

The batteries of stationary storage devices and electric vehicles can serve as a buffer for fluctuating renewable energies, but technical obstacles, high unit prices, and consumer acceptance issues may delay their mass market diffusion. Under which business models is storage already today economically viable, according to market experts, or when will it become profitable beyond subsidies? How likely is it that batteries of electric vehicles will be integrated as bidirectional storage devices?

Stationary storage

Storing electricity beyond laptop batteries and cell phones is more complicated than storing many other forms of energy. Several technologies that allow for large-scale storage have been developed up to a commercial stage, and some of them are in widespread use. One of the most effective and most often implemented technologies is pump storage, where hydroelectric plants are used not only for using the kinetic energy to generate power, but also to pump water from a lower to a higher reservoir in periods when excess electricity is available or particularly cheap. The process is reversible at a fairly high efficiency, so that differences in spot market prices can generate sizable profits. However, pump storage depends on the geographical setting; its use is limited to mountainous or hilly regions. Since the hydroelectric potential in many industrialized countries is largely exploited and new projects often fail because of environmental protection legislation or resistance from local residents, the construction of new pump storage facilities will be limited.

Compressing air and squeezing it into salt caverns or abandoned mines is another technology for larger-scale storage solutions, but it will be possible only where the geological configuration of the ground allows for it. Despite a fairly large potential, it is only in use in a few locations in the world (Swierczynski et al., 2010). Fly wheels and high-temperature storage have not successfully spread into the market because of high costs and technical obstacles, but they may be used in some niche applications. Ultracapacitors and capacitors store energy in electrically charged plates, and not through an electrochemical reaction as with batteries, which allows for a high number of rapid charge and discharge cycles and a higher power density. In stationary applications, they may be used to provide short-term balancing reserves, but have not yet reached full commercial viability (Pike Research, 2011).

Hydrogen, by contrast, may become a widespread, decentralized solution in the future. Wind turbines or photovoltaic cells can provide the electricity to split water into its components, oxygen and hydrogen, via a chemical process called electrolysis. The hydrogen can then be stored and recombined with oxygen with fuel-cell technology. Most recently, the natural gas grid is being discussed as a candidate for storing hydrogen and using it for combustion, so-called power-to-gas technology. First pilot plants like German utility E.ON's Falkenhagen facility are projected to start producing hydrogen by 2013 (see also Dena, 2012).

Storing large amounts of electricity in stationary batteries has been implemented only under specific circumstances, for example in West Berlin during the Cold War, when power supply was a geopolitically sensitive issue that could have been easily used as a strategic instrument by East Germany and its allies. Major reasons for the limited use of batteries are high costs for battery stacks and low energy density under current technologies.

However, with the emergence of decentralized, largely autonomous subsystems, stationary storage devices will gain importance in the coordination of supply and demand.

> As more and more renewables enter the grid, it will be more challenging to even out fluctuating supply. Because an increasing number of houses employ heat pumps and more functions are replaced by electricity, demand will both grow and become more erratic. As a consequence, it will be more challenging, and thus more costly, to balance supply and demand. All this will create demand, and hence additional business models for

storage. Slowly, this is being recognized by the established players. (Voigt, Younicos)

The market potential in Germany for stationary electricity storage devices is likely to reach 40 TWh in 2040. Investments amount to €30 billion by 2030, according to Deutsche Bank Research (Auer and Keil, 2012).

The alternative to grid renewal

If communities aim to achieve full energy autarky, adequate storage systems have to be established. They have to buffer volatile renewable energy output. However, storage will also become more relevant to the future central grid. More specifically, it will face two major challenges: first, substantial but volatile renewable energy inputs in remote locations and a multitude of private producers feeding their locally produced electricity into the low- and medium-voltage grid; second, an ever-increasing network with interconnected electricity markets, significant power flows across thousands of kilometers, and cross-national efforts to coordinate the flows. While the latter renders the system more vulnerable to cyber attacks and instabilities, the former requires extensive balancing between supply and demand.

> Apart from environmental considerations, there is also a security argument in favor of decentralization. The example of Stuxnet[1] has shown that one USB stick is enough to make a turbine run too fast and cause the whole plant operation to collapse. There are similar risks involved in a large grid. (Voigt, Younicos)

Many technically feasible solutions to introduce storage systems have been realized, but for example, the potential for further pump storage locations in Europe is limited; utilities rather revamp and extend their existing reservoirs. Other storage solutions like compressing air or transforming electric energy via electrolysis into chemical energy, namely hydrogen, will have their first tests in pilot plants in the near future.

> If we had storage facilities, at noontime we could dampen supply peaks – for example the photovoltaic peak in the low-voltage grid. But if energy is transmitted to big pump storage plants in the Black Forest, Austria, and Norway, substantial transmission losses occur, and these lines have to be bolstered. However, photovoltaic energy can be stored more efficiently in small, decentralized storage systems like lithium-ion batteries, whereby the production and the point of consumption are nearby. This strategy would reduce the load on the grid and, to some extent, avoid an expansion of the grid and reinforcement of the substations.

> In the case of wind power, we need completely different storage systems, though. Excess electricity could be converted into methane by electrolysis and methanation and then delivered via high-pressure gas pipelines to the major points of consumption, or it could be converted in liquid gas for storage close by the wind power-plants. (Hose, ODR)

The major obstacle of using batteries as buffers for electricity is the high cost. During the Cold War, West Berlin received enormous subsidies to maintain some economic activity. The batteries were largely funded by Western Germany, thus indirectly via the regulated tariffs by electricity consumers. By contrast, a fully liberalized electricity system may not require direct subsidies or financial incentives that promote the deployment of any kind of energy storage. This is because price spikes during peak demand in the wholesale market may create sufficiently attractive business opportunities for energy companies and entrepreneurs to benefit from frequent phases of supply scarcity by installing these systems.

> Just as solar or wind power installations, in the future, battery parks will be financed by funds or private investors. However, such parks will not derive their cash flows from feed-in tariffs for renewable energies but from participating in the grid-balancing market. (Voigt, Younicos)

One of the companies that has pursued this path is Younicos, a company founded in 2008 with headquarters in Berlin. One of the founders and the current CEO of Younicos, Alexander Voigt, explains where batteries are currently cost-competitive:

> At present storage units are still very expensive. In Germany they are commercially viable in limited areas of the energy market, especially in the balancing market. We have still not arrived at attractive prices for house storage units in combination with photovoltaic systems.
> But we are already competitive where prices are high. In many places, such as islands and other remote areas, electricity is generated very expensively with diesel fuel. On islands for example, diesel fuel is not only increasingly expensive because of rising spot-market prices, but also transport costs are high and themselves a function of the diesel price. (Voigt, Younicos)

Electric cars and bicycles, cell phones, laptops, and tablet computers require energy storage devices, thereby creating steep learning and cost-cutting curves in the industry. Voigt claims that the technology for batteries in electricity supply has become financially viable in 2012:

> Power plants solely providing balancing energy on the basis of lithium-ion batteries have already reached economic viability in 2012. That was not the

case in 2011. The number of cycles was still too small and the price was too high. (Voigt, Younicos)

The increasing amount of volatile energy supply imposes a technical burden on the system. Given that energy provision is still considered part of the core infrastructure services offered in basically all countries, governments take the ultimate responsibility of the functioning of the system and have become aware of the necessity to provide a backup system:

> When we started out, nobody was talking about the grid of the future requiring fast controlling units. But the topic has found a broader audience in the energy industry and in the political sphere. There will be changes to existing regulations to define and enable the use of storage units. A great deal can be attributed to activities and operations of our company. (Voigt, Younicos)

On the European level, progress is made more slowly than in national regimes, according to Voigt:

> The renewable energy industry should really be a European issue, but on the European level, time-scales are different from those in the national context. Since we were founded five years ago, there has been no real development in the EU. (Voigt, Younicos)

A large market potential for storage solutions is likely to be tapped soon:

> One of the four major energy suppliers calculated with their own team that it is cheaper to install a home storage unit in every building than to dig up all the streets in cities again in order to expand the distribution grids. (Voigt, Younicos)

However, a connection with the central grid helps to maintain financial viability.

> If one wishes to be completely self-sufficient, the storage units become exorbitantly expensive because they are not needed for 95 to 98 percent of the year. (Voigt, Younicos)

Regional utilities like ODR envisage cooperations and alliances to fund new storage devices within their service area.

> Some projects are a little too ambitious and too capital-intensive for a regional supplier. But in the case of storage systems, it might well be that

ODR, for example, will enter into a cooperation with a battery producer. The battery producer also cooperates very closely with an automobile producer and is doing intensive research, for example, into lithium-ion battery technology. Within the next four to five years, they will offer a high-performance battery. Monetary incentives for local generation will be cut step by step. The funds must go into storage systems. If we do not get the storage units, then the energy transformation will ultimately have failed. Those in government and others bearing political responsibility must think about how to promote these storage systems. (Hose, ODR)

Hose criticizes that massive investments into the central grid could be avoided by extending the storage systems. Once those storage systems are in place, the reinforcement of the existing grid would not make sense any longer. A substantial misallocation of resources may take place due to the wrong regulatory incentives.

At present, the sequence of the steps undertaken to enhance the energy transformation is not properly balanced. We start by strengthening the grids, but in the long term, less and less electricity will be transmitted. We concentrate on expanding decentralized generation plants but do not keep up with storage systems and smart grid systems.

In order to meet statutory requirements, we are now investing in intelligent systems but do not yet have adequate storage systems. When intelligence is all finally in place and we know at any time who is feeding in what and where, and when we have weather forecasts and storage facilities, then we will not need powerful grids any longer. Electricity will remain in the region. The expansion of substations and power lines may be important now, but not in the future. (Hose, ODR)

Developing the blueprint for carbon-free energy systems (Younicos)

Younicos was founded in 2008. Alexander Voigt, the CEO of the company, had been involved in establishing two major German producers of photovoltaic panels, Solon and Q-Cells, but decided to move away from that line of business.

Back in 2006, it already became apparent that a business model with fixed feed-in tariffs would not be sustainable in the long run because the expansion of the renewables would create enormous problems in integrating these fluctuating amounts – at the latest, once the renewables reach 20 percent, which is the figure we have now in Germany. If you are on the board of a company that is based solely on the feed-in tariff market model, and you know that it cannot carry on much longer, then you have to reorient yourself. The German law on renewable energies is part of a command-and-control economy: fixed, guaranteed electricity prices, irrespective of the time of the day or season when feed-in takes place. We have set ourselves the goal of devising a business model that will be needed when up to 80 percent of the energy fed in is generated by renewables. That is not a core business area of a company that produces photovoltaic cells. Thus, we decided to set up a new company. (Voigt, Younicos)

What sets the enterprise apart from many other start-ups in the energy industry is the combination of managerial and technical expertise:

If a company works toward renewable energies and one day becomes the backbone of the electricity supply, one must get to grips with the technical problems that surface on the way. Getting on top of the technical problems to a reasonable degree also means having a command of all the business models that are needed to give economic life to the new technology. Every new technology needs new business models. Younicos is working on both areas. When we set up the business, we were the first to see this field in such an integrated manner.

Our core competence is battery management, understanding batteries and treating them so that they last as long as possible. This is where chemistry encounters economic viability. In addition, we are competent in grid controls, that is, adjusting and stabilizing electricity grids. We are not aware of any other firm that is focusing on what we are doing. Younicos serves as a trigger where new technologies are integrated into the energy industry in a commercially viable way.

We have a staff of around 60. About 35 of them are engineers, especially for mechanical and electrical engineering, and IT. The others are financial analysts, traders, marketing experts, strategists, and sales people who are needed if one wishes to enter a new market with a new technology. One employee is in charge of legal matters, the rest is taken care of by external service providers. (Voigt, Younicos)

As opposed to many manufacturers of photovoltaic cells, Younicos is not a publicly listed company:

We are exclusively financed by equity and have private investors. We are not publicly listed and are completely independent of banks. However, we cooperate with banks for project financing. (Voigt, Younicos)

Requirements for the technical core component of the company's island network, the batteries, differ substantially from the batteries used in transportation:

> The automobile industry also works on batteries. But there the requirements are more challenging: Batteries should be able to withstand temperature differences between −40°C and 80°C. The batteries are shaken, have to function in winter, and must not come whistling through the air in a crash. These are all tasks that have absolutely nothing to do with battery cycles in stationary operation. The automobile industry is keen on batteries that last for around seven years with a very high degree of reliability, but then break down as fast as possible. The industry has no interest in fitting cars with batteries that will last for 20 years. For us, if they are to be economical, batteries must last 20 years without failing. (Voigt, Younicos)

The vision of the company is to devise a viable business model for autonomous, renewable energy systems. Islands provide a natural experimental setting to test the technologies under business conditions:

> We are a business enterprise and not a research institute. If we develop something new, we have to know where we can make sales.
> We are already competitive where prices are high. In many places, such as islands and other remote areas, electricity is generated very expensively with diesel fuel. The energy supplier of the Azores, for example, showed great interest in our concepts. Together, we selected Graciosa, the second-smallest Azores island, for supplying it with 70 to 80 percent renewable energy. The technology and the availability are still very much in their infancy, but we know from our past tests that the system works. Now, the next step is implementing our technological solutions there as an island system. (Voigt, Younicos)

The insights that will emerge from the geographically isolated setting may later be transferred to larger applications:

> The project is on a scale of €25 million. It offers the opportunity of experiencing and condensing the whole learning curve that will be required in Germany over the next 20 years. If we want to have 80 percent renewables here, we will have to work with exactly the same market and technical problems that we are now already dealing with on a small scale in Graciosa. (Voigt, Younicos)

Even if Younicos' application of the technologies may be limited in size, it may contribute to reducing the high costs for batteries:

Instead of first doing research and then later implementing new technologies on a large scale, a modest but quick start could make storage units cheaper sooner. (Voigt, Younicos)

The CEO assumes that the market potential in developing countries is particularly large:

Financial experts and investment bankers scrutinize the market and spot potentials better than politicians or journalists. What we are doing here is an experiment. But when it is up and running, it will also bring us worldwide success.

We have gigantic opportunities because here we have technologies for extremely dynamic markets with soaring energy requirements, above all in the BRIC countries, Brazil, Russia, India and China.

When ideas mature, people from all corners of the planet arrive at fairly similar conclusions. However, we are pioneers because we do not just talk, but everybody can see that our projects work and pay off. (Voigt, Younicos)

The company decided against an early involvement in the North American electricity system:

The United States is also confronted with a new grid architecture due to a high percentage of renewable feed-ins and energy from decentralized generation. We took a technical and advisory part in this process but decided not to enter the market with our own products because, as a relatively small enterprise, we could not cover everything. In the United States there are completely different technical standards – 60 Hertz and 120 Volts. All the technology we have developed here would have had to be duplicated for requirements in the United States. That would have been too challenging for us. As soon as we have moved our enterprise to the next stage, we will also enter the US market. (Voigt, Younicos)

Other storage technologies do not interfere with their business model:

Power-to-gas is not a direct technological competitor for us, since it involves huge chemical plants that have to run very consistently on permanent excess electricity. That may be the case in 2030 or 2040. Until then the focus will be on short-term balancing of energy provision. (Voigt, Younicos)

Local utilities may enter the storage market through cooperations and alliances, according to Frank Hose of regional grid operator ODR:

For ODR or EnBW, it is appropriate to seek cooperation with a storage unit producer before any storage unit producer enters the market. They supply storage units; we sell the storage system and the smart meter to the customer and implement demand-side measures. (Hose, ODR)

Electric vehicles

A hype revisited

The increase in the amount of fluctuating electricity from renewable energy sources, especially wind and sun, leads to a greater need for the temporary storage of generated energy levels and associated network services than before. One of the decentralized storage systems available for the near future is the battery of an electric car. Owners of those cars can actively participate in the electricity market. In comparison with other storage technologies, such as pumped storage power plants, compressed air storage, and the gas grid, batteries of electric cars have only a limited storage potential. Once car owners connect their vehicles to the grid, they can provide so-called ancillary services to maintain the reliability of the power supply by participating in the secondary market of minute reserves, though. This allows electric cars to positively contribute to the stabilization of the network and facilitates the transition to a low-emission power supply in a market with a high share of renewable energy. Electric vehicles may also serve as batteries for autonomously produced electricity. For example, commuting residents who own photovoltaic panels can charge their vehicles for small distances or install complementary stationary batteries to store electricity produced during daytime for recharging of the car at night.

Ullrich Müller leads the corporate strategy department Global Regulatory and Innovation Strategy at car manufacturer Daimler. Among other tasks, his assignment encompasses worldwide responsibility for research strategy, strategy on sustainable products, and regulation on all automotive issues worldwide. He explains the link between electric vehicles and the electricity grid:

> Electric vehicles can become an important component of the grid. The batteries may discharge when electricity is needed or charge when excess power is available. This grid balancing would have positive effects on operating costs and on environmental impacts. Therefore, there is a close link between electro-mobility and the smart grid.
>
> The storage facility of an electric vehicle is the battery. The concept of batteries was found by Alessandro Volta more than 200 years ago, and we use state-of-the-art lithium-ion batteries in multiple household appliances like cell phones.

The main challenges for a decentralized energy and mobility revolution are storage technologies and cost efficiency. The lithium-ion battery is the only commercial technology currently available that allows a reasonable driving range and workable size for electric cars. But the real breakthrough will only be achieved when lithium-air or lithium-oxygen batteries are ready for commercialization – these technologies will not be broadly available in the next decade. They will be capable of storing four to six times more energy, which would substantially reduce the size of the battery, boosting the range beyond the current standard of around 150 km. The smaller vehicles, mostly in urban areas, could then be solely fueled by electricity from batteries, and the larger ones would more likely be plug-in hybrids, running on electricity for distances around 50 km and then switching to an internal combustion engine for journeys beyond that range. On a large scale, this scenario is not likely to happen before 2020, though. (Müller, Daimler)

The costs of currently available batteries increase the price of an electric vehicle substantially. However, within the next decade Müller predicts a steep cost decline:

At the moment, the cost of a battery is close to €1,000 per kWh of storage capacity. A cost reduction down to €250 is expected between 2015 and 2020, when economies of scale set in. The technology only becomes economical when mass production is reached. Until then the total cost of ownership will be much higher than for a conventional vehicle. (Müller, Daimler)

The diffusion of electric vehicles depends on a multitude of atomized vehicle purchasers who individually decide whether to buy or not to buy an electric vehicle, and in the second step – if they decide to buy one – whether to use the battery of their vehicles to provide grid services.

The major competitor of electric vehicles is the internal combustion engine, typically based on fossil energy resources like gasoline, diesel, or natural gas. The chemical composition of these fuels allows them to act as highly flexible storage devices and reservoirs. Compared to the electrochemical capabilities of current lithium-ion batteries, fossil fuels have an energy density, that is, the energy available per mass, approximately 15 times higher than the chemical energy stored in batteries (Anthony, 2012).

Hence, vehicles with conventional, combustion-based power trains are much lighter than batteries and can be more easily used for transportation. According to Daimler, conventional combustion engines will continue to dominate the market for individual transport in the future.

The challenge is to continue developing not only electric and fuel cell vehicles, but also established and improved conventional propulsion technologies. If only 1 million electric cars are part of the German vehicle stock in 2020 and the total number of vehicles on the roads is around 40 million, then it is clear that conventional powertrains will carry the main burden of mobility also in future. The classic combustion engines will become more sophisticated, while being complemented by alternative technologies. (Müller, Daimler)

Apart from strong competition of vehicles with an internal combustion engine, battery technology has a number of drawbacks.

Batteries are still adversely affected by quick charging. The more often a battery is recharged, the shorter is its usable lifespan. Customers would therefore have to be given a major financial incentive to accept the risk of their batteries becoming dysfunctional earlier than is specified or acceptable. (Müller, Daimler)

According to Alexander Voigt of island solutions developer Younicos, the battery technology for electric vehicles is not yet sufficiently advanced to allow for additional battery cycles in support of the balancing market.

At present we do not consider that electric vehicles can also be used for storage or for balancing energy. Through our close contacts with the automobile industry, we know that it is not an easy task to reach an adequate number of battery cycles even when the car is just used for driving. If an extra burden is placed on the car through balancing energy with additional cycles, a problem arises for the manufacturer with regard to the technical guarantee. So, this might be a solution 50 years from now but not before. (Voigt, Younicos)

The technological context in which electric vehicles are deployed is characterized by an increasing convergence of the electricity sector with internet and communication technologies. The required coordination of the charging behavior of car holders coincides with the plan to create an intelligent distribution network. Synergies may arise from the combination of different technologies with the help of appropriate policy instruments. But even under favorable policy scenarios, overall participation of electric vehicles in the electricity system would be marginal in the foreseeable future.

Even a million electric vehicles would only account for 1 percent of the demand for electricity. That will hardly provide a robust backup for the public grid. (Müller, Daimler)

Müller differentiates between different electric vehicle technologies, depending on specific traveling and transport requirements.

The search is ongoing for new future possibilities, not only for passenger vehicles but also for commercial vehicles conveying both persons and goods over short and long distances. In the long-term planning of transport routes, hubs should be created outside towns and urban agglomerations. These hubs will be the point of transfer to vans or medium-sized trucks that serve the downtown areas partially emission-free. The issue of electro-mobility will therefore take on particular importance for small commercial vans.

In contrast, one can hardly imagine heavy trucks being powered by electric batteries. The costs and weight of the battery would simply be too high, the performance and the range would be too low. The hybridization of a long-distance truck could result in a cut in CO_2 emission of 4–6 percent; this is the state of technology today. In the case of a medium-sized lorry, mainly employed in stop-and-go traffic, a mild hybrid option is sufficient to be able to travel 1 to 3 kilometers electrically when entering a town.

In public transport, we are already registering strong demand for hybrid buses. There will also be hydrogen buses, which we have already tested with a subsidiary in 12 cities worldwide. Fuel refilling infrastructure can be installed at the depots for refueling to take place at night or whenever they return to the depot. Consideration must also be given to whether city buses running on batteries have a role to play, if, for example, recharging- and quick-charging facilities are installed at coach terminals. (Müller, Daimler)

Increased competition may emerge from Chinese car producers.

From our cooperations we can see how committed the Chinese are to electro-mobility. China wants to decrease their dependence on imports of crude oil. Already years ago, they announced that they want to move into electric battery technology via the hybridization of petrol-driven vehicles. They want to prepare the next technological leap forward. For more than five years, they have not invested in other forms of propulsion but in electric battery drive technologies and hydrogen fuel cell vehicles. The conventional combustion engine is only accepted there because hybridization already reduces fuel dependence.

Electricity for electric cars can be generated from different primary resources. The Chinese have enormous reserves of coal, which they will naturally use, both for generating power to fuel electric vehicles and to start on hydrogen technology. In the future, they are likely to threaten the position of German car manufacturers. (Müller, Daimler)

Apart from cars with an internal combustion engine, several alternative powertrain technologies may also threaten the mass-market success of electric vehicles. They include biofuel and natural gas (LPG-CNG) combustion engines, and fuel cell vehicles. According to Müller, vehicles with a powertrain based on a fuel cell can be expected to be economically viable soon.

> Working in consortia, we have a positive business case when building up an infrastructure for fuel cell vehicles. Today, we do not yet see that for electric cars powered by batteries. (Müller, Daimler)

While Daimler is testing fuel cell vehicles in parallel to electric vehicles, utility GASAG is promoting natural gas-powered vehicles:

> Electro-mobility will find its place in the market, for example in car-sharing schemes, but initially it will be limited to short-range journeys. At first, it will not become a direct competitor for a gas-powered vehicle. The natural gas car has to prove itself against petrol and diesel vehicles. VW offers a Passat run by natural gas, which is suitable for everyday use and provides an absolutely satisfactory performance, even for demanding customers. More manufacturers will have to be motivated to broaden the range of vehicles they construct. By 2015, manufacturers in the European Union will have to meet challenging corporate average fleet emission targets. Natural gas vehicles can complement the range of models so that these targets can be achieved. Electric vehicles are not yet so advanced.
>
> We are one of the pioneers of natural gas vehicles. In Berlin, we launched the "1,000 environmental taxis" scheme. In contrast to the electric car, the natural gas car is already available today and its driving characteristics can be compared with conventional petrol and diesel vehicles. Customers looking for alternative, individual mobility without paying more or sacrificing comfort ought to invest in a natural gas car. (Prohl, GASAG)

Combining lead markets and lead suppliers (Daimler)

Electric vehicles were a serious competitor to cars with internal combustion engine – 100 years ago. Several attempts to re-establish the technology in the second half of the 20th century have failed.

> Research into the issue of electro-mobility has been ongoing for around 20 years. Daimler has already investigated all conceivable options. Since hybridization also requires an electric drive, we have been working on batteries and electric vehicles in parallel to the development of hybrid vehicles. But it is only in about the last five years that the budgets have been high enough to achieve automotive quality standards and corresponding reliability. (Müller, Daimler)

Daimler aims to reach the state of mass production of electric vehicles in the next years.

> We are now well past the test phase. Some pilot projects are running and we are embarking on the production and sale of electric vehicles. Market launch and mass series production are on our agenda because research and pre-development are largely completed. We will be launching the Smart with the lithium-ion battery, and in the coming years the S class as a plug-in hybrid. All alternative powertrain vehicles that we have built so far are not yet being manufactured in series production, but we intend to increase daily production substantially. The S-400 Hybrid was for us a very important vehicle to prepare series production for future launch of alternative powertrains. Especially the handling of high-voltage components in factories and services is now well understood. (Müller, Daimler)

Compared to other car manufacturers that already have electric vehicles in their sales portfolios in 2012, Daimler is confident that it will not suffer due to its late arrival on the market.

> Daimler does not lag behind other car manufacturers. For example, technological development of the E-Smart is already definitely competitive. As a commercial enterprise, you always have to present a viable business case – I do not see that with many of our competitors at the moment. (Müller, Daimler)

Even though battery stacks and electric motors can be considered commodities in a global market, Daimler has decided to establish cooperations with battery manufacturers. This strategy facilitates the possibility to respond to specifically tailored requirements of individual vehicles.

> A battery for a hybrid vehicle, for example, must have a very high performance and supply a lot of energy in a short time in order to bring the torque into play. It is not easy to find products in the market that meet our requirements with respect to safety, reliability, performance, size, temperature, charging cycles, etc. We anticipate gaining a competitive advantage if we are in the game right from the start, but we do not wish to embark on cell production – we have to integrate batteries in the vehicles and they all have different requirements, depending on their class and the model. That is why it makes sense to go along with the development in the early years and to overcome obstacles in technological development. Moreover, it is in our interest to make a clear reduction in the size of the battery in order either to

> accommodate more battery in the car or to save space. Durability, certification ability, etc., equally have a part to play. That is why we have to make a stronger commitment in the early stages.
>
> Of course, both electric motors and lithium-ion batteries are well known. But a vehicle requires different specifications than, say, cell phones or laptops. That is why we have entered into cooperations to get involved in the development of batteries and electric motors and to bring in automotive know-how. (Müller, Daimler)

Inside the corporation, the units in charge of electric vehicles have been bundled.

> In the vehicle research department, we have concentrated all the relevant propulsion and development units for electric propulsion. This also includes predevelopment. We have created a unit that is in charge of electro-mobility as a whole. It has the task to coordinate the whole product cycle from the development of the powertrain over mass production to global service and aftersales. (Müller, Daimler)

While the use of electric cars in the network has to be organized and coordinated, the individual car owners have to get appropriate incentives to connect their cars wherever possible with the network in order to obtain maximum benefits from their services.

Despite general skepticism vis-à-vis regulatory interventions, Daimler sees the necessity for an adequate regulatory framework that promotes electric vehicles.

> Regulation is a tricky topic. Generally speaking, we do not want regulated markets, but competition. However, it requires huge investments to quickly build up production capacities and move from the pilot phase to mass series production. Companies expect some state support to achieve that. At least, planning certainty should be created for companies that are now redrawing their investment plans in order to make a concrete start on the production of batteries and electric motors. Making vehicles more attractive for customers could be achieved, for example, by incentives to grant free parking or permission to use bus lanes. Of course, infrastructure for charging batteries is also required. If Germany aims to become a lead market, we need stimuli that offset major cost disadvantages of electric vehicles. Admittedly, no one can predict whether by 2020 we will have a million vehicles or a great deal less. But if we want to reach a million by 2020, there will have to be enormous growth on the electro-mobility market.

> From the angle of our basic political philosophy, it may appear sensible to refrain from distorting subsidies, but as in the international context, competition is being shaped by national industrial policies. If the right overall conditions are created more speedily and more successfully in other markets rather than in Germany, then the sales charts for electric vehicles will record a soaring curve in those countries because ultimately electric vehicles are sold where there is demand. (Müller, Daimler)

A governmental purchase incentive would accelerate the market diffusion of electric vehicles. The sales of electric vehicles are likely to first occur in countries with more generous government subsidies.

> The customers are to be found where the best overall conditions predominate and where purchase incentives and discounts are implemented. These include, for example, charging stations or state programs such as in France, where electric vehicle purchasers receive a subsidy of €5,000. In particular for smaller cars, this is an absolutely crucial factor in the purchase decision. (Müller, Daimler)

Municipalities also have some means to positively influence the local diffusion of electric vehicles.

> German cities have already begun to integrate not only operating costs but also CO_2 abatement costs in their tenders for public transportation. The CO_2 emitted when the buses are running is set against the costs for avoiding them elsewhere. If topics such as motorway tolls or city center congestion charges are implemented, zero-tailpipe emission vehicles become more attractive. (Müller, Daimler)

Pure battery-electric vehicles require an infrastructure that allows for quick charging while on the road – with a conventional plug these car holders would have to cope with long waiting times until their batteries would be fully charged. The lack of a network of quick-charge stations creates a large impediment for car owners who frequently travel longer distances. Initial quick-charge stations are currently under construction, but high installation costs and immature technology to optimize the interaction between battery chemistry and quick-charge stations are delaying their rapid spread.

> The infrastructure for electro-mobility is relatively dispersed, as we have too few charging stations. Of course, the vehicle can be charged at home if there are no time-constraints. But public and semi-public charging stations in the cities – for instance in customer parking lots, where a battery

can be recharged within one hour – are important for responding to the limited range in urban transport. These quick-charge stations presently cost between €7,000 and €9,000. Of course, there will be more charging stations if prices fall in the next few years. Other countries such as China and the United States are already vigorously expanding this infrastructure, though. (Müller, Daimler)

Some municipalities are in the race to the top, though, and intend to install charging stations on a voluntary basis.

We will make a contribution with regard to electro-mobility. Around 20 percent of households have a second or even a third car. Here we are trying to boost the share of electric cars by offering charging points at home for free together with an all-inclusive package for recharging. Initially that costs us money but it is necessary to change people's frames of mind, to change mindsets. That works best via the price. And we attach more customers to us. (Jänig, SWU)

A similar type of regulatory cherry-picking is suggested by Eckhardt Rümmler of energy incumbent E.ON. Multinational companies like E.ON or Daimler operate in various markets and can build their strategies' regulatory specificities more efficiently than players in a domestic or regional setting. By exploiting opportunities established by environmental policies, those companies can serve as channels and triggers to accelerate the learning curve and become dissemination leaders.

We analyze the regulatory situation and natural conditions in all countries where we operate and what could fit in with our plans. If we had wanted to invest massively into highly subsidized photovoltaics in a country with rather average to below-average natural conditions for solar, we would have done that in Germany. In England CHP is attractive, in the Czech Republic electro-mobility, in Sweden biomass and biogas. We consider which technologies of our portfolio can be developed to market maturity and then take them into the country where we find the best regulatory and natural conditions. If the product has reached market maturity, we re-import it.
This makes us not really an innovator but a smart follower. We do not wish to manufacture individual components and hardware. But we develop applications, for example for smart homes, and search continuously for new promising business models linked to customer needs, technological developments, and changes in market design. (Rümmler, E.ON)

Findings on stationary storage and electric vehicles

- *Stationary storage devices will soon become economically viable in secondary electricity markets and in isolated grid settings*: The future energy grid will consist of loosely connected island systems with a high degree of autonomy and less frequent power exchanges than today. Stationary battery stacks will become indispensable tools to balance fluctuating supply and inelastic demand within these areas. Companies with capabilities and expertise in systemic management of these components will play a key role in the energy system of the future. Under current cost conditions, though, batteries are only profitable in grid-stabilization services and remote locations with a sufficiently high intake from fluctuating renewable energies. Companies can gain experience with control and integration of storage devices into the grid by exploiting these opportunities.
- *Electric vehicles will not be a key element in grid stabilization but as storage devices for individual households*: Batteries in electric vehicles are not likely to substantially contribute to balancing the electricity system in the foreseeable future because the shift from cars with an internal combustion engine to electric vehicles will progress very slowly and will capture only a small share of the overall stock. In addition, the technical configuration of batteries for electric vehicles is not favorable to substantially increasing the number of charging cycles. However, individual usage as a proprietary buffer for privately produced electricity is an economically attractive option for home owners with photovoltaic installations, once that feed-in subsidies cease.
- *Electric drivetrains will occupy niches in municipal services and industry*: Local environmental regulation, for example restrictions on access to city centers for zero-emission vehicles, may accelerate the market diffusion of vehicles with electric or hybrid powertrains. In particular, local delivery and industry and public transport services may rely on small commercial vans and electric buses, respectively, because of the relatively small distances they cater for.
- *Multi-country presence drives learning processes and creates momentum*: Governments engage in a race to the top by fostering and promoting lead markets. They use different policy incentives, like purchase incentives for electric vehicles or investments in the charging infrastructure. Companies operating in multiple markets can cherry-pick and launch new products and services in countries that offer the best regulatory environment for their innovations. By exploiting these incentives, they allow for spillover effects and positive externalities for their product development, even in countries with less favorable conditions.

6
Enabling Negawatts

As outlined in this book's introduction, efficiency improvements in the buildings sector should be considered an integral and important part of decentralized energy services – not only because of the fact that conservation measures often coincide with the installation of a combined heat and power unit. Given its enormous savings potential with respect to reducing greenhouse gas emissions, the buildings sector deserves special attention among the various dimensions of the transformation of the energy system. Technologies have advanced, and green buildings are nowadays a unique selling proposition for architects and investors. However, improving the building stock – especially in industrialized countries with modest new construction activities – is proving to be a major challenge, not least because of the atomized ownership structure.

Which business models reap the benefits of the vast efficiency potential and can overcome problems related to split incentives and incomplete contracts?

Theoretical framing

Measures to increase energy efficiency are the most cost-effective way to proceed toward a low-carbon economy – many even have *negative* costs, which implies that they result in a net profit for companies and investors. On the marginal abatement-cost curve regularly released by consulting practice McKinsey, investments in efficiency offer the greatest potential to curb carbon dioxide emissions for the least cost.

As McKinsey analysts write in a study on abatement potentials in Germany: "In the buildings sector, levers to reduce energy consumption and increase energy efficiency such as insulation, replacement of heating systems, facility management systems, efficient electrical devices and lighting, contribute most to greenhouse gas abatement…. As the additional investments required for these levers often lead to high energy savings, almost 90 percent of the abatement levers (63 Mt CO_2) in the buildings sector pay off for the decision-makers within the respective amortization period" (2007, 19).

Figure 6.1 on the following page, depicts the marginal abatement-cost curve for the German buildings sector. All measures shaded in grey and below the horizontal axis have negative abatement costs, that means that saving a ton of CO_2 is actually profitable for the investor.

But if all those measures involve negative costs, why are they not implemented? Is it a lack of adequate regulation? May it be linked to the fact that other, much more expensive policy measures are more attractive to politicians? Why would, for instance, photovoltaic energy receive generous feed-in subsidies – even though energy savings in buildings are much more efficient – if the overarching objective is to reduce greenhouse gases?

The vast potential of energy efficiency activities has been in the public debate for the past few decades. In the ground-breaking manifesto "Factor 4," published in 1998 by Amory and Hunter Lovins of the Rocky Mountains Institute, and by Ernst Ulrich von Weizsäcker, the former director of the German Wuppertal Institute, the authors suggest that energy consumption and economic growth can be effectively decoupled – "doubling wealth, halving resource use" (von Weizsäcker et al., 1998). More than a decade later, their aspirations have not changed, but the world apparently has not taken sufficient initiatives to implement their fertile ideas. Rebound effects may account for a limited share of the disappointing performance of industrialized nations, but the conundrum persists – namely: Why do the apparent benefits of energy efficiency not ignite broader action? This phenomenon is commonly known as the "efficiency gap."

For the buildings sector, a simple answer does not exist. It rather seems like an imperfect blend of half-hearted regulatory efforts and financing constraints leading to real market barriers, path-dependence aspects of durable goods, and the principal-agent problem, which may explain the absence of efficient measures to increase efficiency. As the McKinsey analysts write, "implementing these abatement levers often requires overcoming substantial obstacles. These include the total investment needed, relatively long amortization periods of more than 10 years, and the unequal distribution of costs and benefits of a measure (e.g., between tenants and property owners)" (2007, 20).

The following sections will be dedicated to disentangling the efficiency puzzle. For that objective, we want to distinguish between *market barriers* and *market failures*. While the next section deals with market barriers, the two subsequent sections tackle the most prominent market failures.

152 *The Decentralized Energy Revolution*

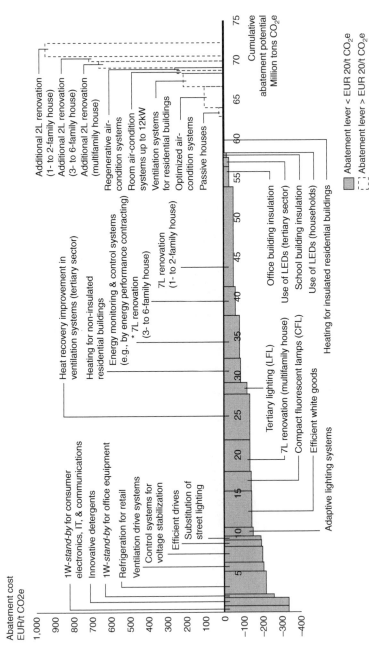

Figure 6.1 McKinsey abatement-cost curve for the buildings sector in Germany, 2020

* 7L renovation is a renovation which reduces the heating needs of a house to 7 liters of heating oil per square meter every year.

Source: McKinsey (2007).

Inelastic demand for durable goods

A market barrier can be understood as an obstacle that prevents specific policies from being implemented or investments from being undertaken. One market barrier in energy efficiency is that it traditionally had a lower policy priority for politicians than other, more fashionable and fancier incentives. From a bottom–up perspective, a market barrier exists when residential consumers in industrialized countries do not consider an additional investment in an energy-saving light bulb one of their key priorities. This is because individual savings may be negligible compared to the total available budget, even if – considering society in its entirety – the sum of all savings may be substantial. A further barrier for energy efficiency is worth mentioning: Often energy efficiency features, for example of a newly planned office building, are only a minor aspect of the overall design, including aesthetics, space functionality, etc. Energy efficiency is part of a bundle or package of features and is not provided as an independent product – it is hence considered an "incomplete market."

While these obstacles can be overcome by increasing pressure on politicians and adequate end-user and building regulations – as implemented recently with energy-saving light bulbs in the EU – one largely unresolved market barrier in energy efficiency is access to capital. The finance costs for residential and small commercial consumers may be high, and even in industry, internal hurdle rates for energy efficiency investments may be higher than the cost of capital to the firm (DeCanio, 1993).

Private and public investors are also deterred from taking action because of a high degree of perceived uncertainty related to energy prices, although reducing overall energy consumption can be considered an effective tool for hedging in markets with finite resources. Even if energy-efficient products are more expensive than their conventional counterparts, when all costs are discounted and added over the lifetime of a product, energy-efficient products are most likely to result in actual cost savings. But the purchase decision may not reflect pure monetary rationality. In an experiment by Meier and Whittier (1983), as quoted by Brown (2001), people had to choose between efficient and less-efficient, but cheaper, refrigerators. Despite apparent monetary savings over the longer term, more than 50 percent of the participants chose a cheaper refrigerator.

From an economic point of view, buildings equipment can be categorized as "durable goods." An investment into such a durable good typically involves a sizable portion of loan capital, and private investors may hesitate to take that financial risk unless some outside trigger initiates action. For example, the replacement of heating systems in private homes does not correlate with government incentives: Homeowners purchase a new heating system when the old one becomes dysfunctional. Data on that issue from Germany show astonishingly little effect of efficiency incentives launched by the government: The rate of purchases of new heating systems fluctuates fairly arbitrarily, irrespective of new legislation. Figure 6.2 depicts that phenomenon for the time period 1990 to 2008.

Only in 1995 did a peak in new heating systems installed coincide with a change in regulation, whereas the other significant legislative step in 2002 did not generate any visible consequences. Johannes Hengstenberg, founder of web-based consumer advice platform co2online, explains why the modernization of buildings remains fairly inelastic:

> Around 1.4 percent of all German buildings get new insulation each year. For heating systems, the proportion is just above 3 percent, with a decreasing trend. Hence, heating systems in Germany are aging. The modernization of pumps is also stagnating. Replacing old systems accounts for 90 to 95 percent; there is hardly any proactive modernization before old pumps become dysfunctional. Yet pumps and heating boilers are worthwhile investments; they are more secure and pay better interest than depositing the money at the bank. (Hengstenberg, co2online)

Hengstenberg comments that the market for residential homeowners is tricky and too phlegmatic for new business ideas, including two that are discussed in this book.

> Many companies with new ideas fail in the residential housing sector. Kofler Energies, for example, entered the residential housing market with the wrong strategy. The initial announcements by LichtBlick were also too

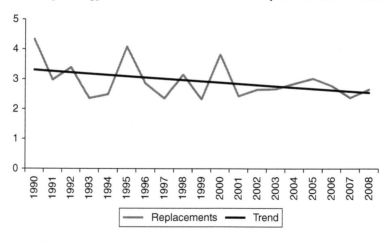

Figure 6.2 Share of German heating systems replaced annually
Source: co2online (2012).

optimistic. The number of CHP units they hope to sell will certainly have to be corrected downward. (Hengstenberg, co2online)

In a report to the US Environmental Protection Agency, the US National Association of Energy Services Companies identifies three specific market barriers (ICF, 2007) – two of them related to financing constraints: First, the well-developed commercial real estate industry in the United States that owns buildings and leases them to tenants, hesitates to undertake any long-term debt obligations that might prevent them from reselling their properties on short notice. Second, many US manufacturing companies avoid long-term commitments with project payback requirements of more than two years, which under normal circumstances precludes comprehensive energy efficiency projects. The third obstacle is related to organizational routines.[1]

Numerous government agencies all around the world promote energy efficiency investments, also for residential consumers, like the US "Energy Star" program that until the end of 2011 provided tax credits for energy improvements to existing homes, including water heaters, central air conditioners, building insulation, and new windows and doors. But households may not always be aware of the options they have. In a survey among over 1,000 German house owners who undertook a renovation, 73 percent expressed that they were "not so well-informed" or "not at all" informed about funding possibilities. Even in the sub-sample of almost 550 house owners who implemented a renovation exceeding regular efficiency standards, 60 percent did not feel adequately informed (Stieß et al., 2010, 42).

It seems likely that informational hurdles will gradually be overcome through the internet revolution. The new generation of digital natives is used for accessing information at greater ease than older generations, which still represent the bulk of homeowners. Specialized websites will provide the necessary information on how to apply for government loans and which efficiency measures will yield the greatest benefits. For example, the German not-for-profit company co2online, which is mainly sponsored by the environmental ministry and other public entities, runs several websites that allow for estimates of the savings potentials of private homeowners and connects them to selected local technicians.

Split incentives and the principal-agent dilemma

When two parties involved in a contract overtly or implicitly pursue diverging goals and have different levels of information, the principal-agent dilemma looms. Buildings whose owners are not tenants are a classic case where the dilemma can emerge: Any improvements in the efficiency of the building mean additional expenses to the owner, but often result in a benefit to the tenant.

For instance, a better insulation reduces heating costs for the tenant, but the owner does not necessarily have any financial advantage from the improvement. Conversely, the landlord of a furbished apartment may choose to install a cheap but energy-consuming refrigerator, which generates higher electricity bills for the tenant than a more expensive, less environmentally harmful product.

In economic theory, that phenomenon is called "split incentives," and according to the International Energy Agency it is one of the major hurdles in overcoming the efficiency gap (2007, 12).

In its analysis, the IEA observes that "principal-agent problems are pervasive, dispersed and complex and, as such, no single policy will fully overcome this market failure." In addition, the success of policies depends on the cultural setting and existing awareness of the value of energy savings in a country. The authors comment that energy-efficiency programs in Norway had greater difficulties in getting accepted than in Japan because awareness of energy conservation issues in Japan has traditionally been higher due to the absence of significant domestic energy resources (IEA, 2007, 196). However, the authors identify three key elements that help to reduce their impact.

First, price signals should be integrated more explicitly in any form of arrangement between tenants and energy users as well as landlords and appliance purchasers, respectively. Redesigned contracts may help to "break the isolation of both principal and agents from the energy price signal" (IEA, 2007, 193).

Second, best-practice regulation should be imitated and implemented. In Germany, for example, prices for rented apartments are often capped (although adjusted to inflation, etc.) by municipalities as a means to protect tenants. The mid-size town of Darmstadt introduced a pilot project called "Ökologischer Mietspiegel," which took energy conditions of buildings into account and allowed for an increase of €0.37 per square meter in the rent if the building had undergone an energetic modernization, thus benefitting both landlords and tenants (IWU, 2007).

Governments should promote easily accessible information about energy efficiency in order to raise awareness both with principals and agents.

Transaction costs and incomplete contracts

Often owners of buildings, in particular public entities like schools and municipal administrations, have neither the financial capabilities nor the expertise to renovate and modernize their properties. A large potential of energy savings remains untapped. But private investors may step in and make offers that the owners raise energy efficiency up to latest standards, by outright exchanging the heating or cooling system as well as by operating the facilities more intelligently. Most of these arrangements foresee that the owner will transfer the difference between the theoretical expenses under the old state of the building and the actual costs after the conservation activities to the investor over a certain amount of years. In theory, this concept is a classic win-win opportunity: While the owner gets a modernized building at no extra expense, the investor receives, if calculations are correct, a highly predictable return over the agreed period of time.

Despite its apparent appeal, energy performance contracting for buildings has not been overly successful during the past decades. The theory of Incomplete Contracts may provide an explanation for its lack of popularity and outright market failure. The theory is based on the more general field of Transaction Cost Economics (Williamson, 1985) and was formulated by Grossman and Hart (1986), Tirole (1986), and Hart and Moore (1990) to provide an alternative explanation to neoclassical approaches as to why and under which conditions firms exist and where the boundaries of firms are located. The theory's core concept can also be applied to building and efficiency contracting. Hart's basic idea is that no contract can be complete because the involved parties cannot *think* of all contingencies that may arise, it is hard for them "to *negotiate* about these plans, not least because they have to find a common language to describe states of the world and actions with respect to which prior experience may not provide much of a guide," (Hart, 1995, 23, emphasis added) and eventually it may be difficult to *write* the contract in a way that an outside authority is able to figure its details out and enforce them.

Every building has its unique features, which renders any form of efficiency contracting a highly complex task for all involved parties. In particular, ex-post negotiations are likely. Even if they are anticipated during the project phase, they may cause disputes and raise transaction costs: Who would pay for the replacement of devices indirectly linked to the functioning of the heating system? Who is responsible for leakages or cleaning of the tube system? What happens if overall savings are significantly lower than the contractor calculated? What happens if energy prices evolve in a different way than projected in the reference scenario?

A consulting practice specialized in efficiency contracting reports, for example, of a case in which a German municipal utility issued a tender for 13 public properties. While electricity savings almost reached the target values, natural gas consumption declined only 31 percent from what had been agreed. Two reasons for the underachievement were identified: First, the savings quotas were too optimistic; second, energy conservation in heating requires at least two years for its full potential to be realized (Roß, 2004).

In addition, markets for energy performance contracting are often not sufficiently developed and opaque for outsiders because services are very specialized and contextual and a proper comparison among peers proves difficult. In most countries, the industry has some dominant players, often control or equipment manufacturers and a relatively high share of fringe companies. For example in the United States, eight companies accounted for 79 percent of industry activity in energy service contracting, whereas 61 percent of the companies operating in the field comprised only 21 percent of revenues (ICF, 2007, 13, data for 2006).

Building efficiency

Technological progress driven by standards?

Increasing the energy efficiency of buildings involves a range of technologies, including better insulation, improvements in the heating and cooling system, and smart appliances. A decomposition of the energy consumption of residential dwellings shows striking differences across countries and continents: While space heating is most pronounced in Europe and the United States, households in Japan allocate most of their energy consumption to warming up water, and cooking is by far the largest energy consumption factor in Indian households. By contrast, air-conditioning is virtually absent in countries and regions like China, Japan, and Europe.

Commercial buildings represent a significant share of the total building stock, but building owners and operators are only slowly tapping into that market potential. The International Facility Management Association (IFMA) and the Institute for Building Efficiency conduct an annual global survey among more than 3,500 building owners and operators. Figure 6.3 indicates the most popular measures among the sample building owners and operators.

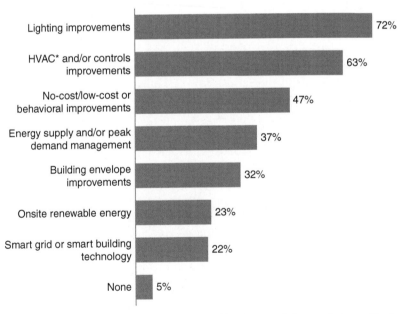

*Heating, ventilating, and air-conditioning

Figure 6.3 Energy efficiency measures implemented in commercial buildings
Source: IFMA/JCI (2010).

The results of the survey show that operators and owners of larger facilities tend to implement the "low-hanging fruits," that is, lighting improvements and changes in heating, ventilating, and air-conditioning, whereas installing onsite renewable energies and smart grid or smart building technologies were not as popular.

> Stand-by switches and light bulbs are the classic appliances to save energy in private households. Unfortunately, the electricity consumption of light bulbs is not particularly high. The use of variable-speed drives in pumps, especially heating pumps in households, is more interesting. In industry, variable-speed drives are a great lever. (Weinhold, Siemens)

German utility GASAG demonstrates the possibilities of building efficiency in their corporate headquarters.

> In our new building we can show how energy can be handled efficiently. Our consumption figures are substantially lower than in our old administration building. We have decreased heat consumption by 50 percent and electricity consumption by 33 percent. The opportunities for cutting electricity consumption were limited since with IT technology we had already reached the latest state-of-the-art, and the IT share continues to grow. In the building, we installed our own power plant that generates not only electricity and heat, but also cools the building in summer. We have tried to reduce consumption for cooling, for example by reducing the window surface area. With this building, we act as a role model and at the same time demonstrate our high performance. (Prohl, GASAG)

The overall objective of innovations in heating and the hot water system is to establish a so-called passive house, which is a house without conventional heating and cooling systems. Such buildings are passive, "because the predominant part of their heat requirement is supplied from 'passive' sources, e.g., sun exposure and waste heat of persons and technical devices. The heat still required can be delivered to rooms by the controlled ventilation system with heat recovery" (Auraplan, 2009).

> If a building is thermally insulated, it only needs a fraction of normal heating requirements. It is no longer worth connecting it to the local natural gas grid unless it has a mini CHP unit. During daytime, when electricity prices are low, an electric heating system can be recharged, while the building itself acts like a storage unit. (Hose, ODR)

While it is technically almost impossible to retrofit existing buildings to become passive houses, advances in the architecture of new buildings – including façade as well as heating and cooling solutions, smart lighting features, and room design – allow for a new generation of buildings that use much less energy but are no more expensive than conventional buildings. However, it is necessary to integrate the energy system into the holistic planning process, as the IPCC notes: "The systems approach in turn requires an integrated design process, in which the building performance is optimized through an iterative process that involves all members of the design team from the beginning" (as quoted in Weizsäcker et al., 2009).

The World Business Council for Sustainable Development estimates that energy savings in the buildings sector could be as high as the current consumption of the entire transport sector, or in total a 60 percent reduction of current energy use in residential buildings (2009, 52).

> Germany spends roughly as much money on boosting the energy efficiency of buildings as on heating fuel consumption, around €30 to €40 billion per year. There should be a switch in the proportions so that only half as much is spent on fuel and correspondingly more on renovation. That would imply extra annual sales of €20 billion for the related industries and crafts firms. (Hengstenberg, co2online)

The Buildings Performance Institute Europe estimates that renovation rates (other than those relating to single energy saving measures) in Europe hover between around 0.5 and 2.5 percent of the building stock per year. By 2050, renovation activities in the baseline scenario could lead to annual energy savings of 9 percent, and in the most ambitious policy scenario up to 71 percent. In all of their scenarios, net savings are possible and range from €23 billion to €474 billion, indicating the sheer market size for this region of the world (BPIE, 2011).

Given their low (or negative) abatement costs, energy efficiency improvements offer enormous market potential. Governments start large-scale financial initiatives to incentivize private homeowners to implement efficiency measures. For example, the German development bank KfW issued an unprecedented amount of €9 billion in subsidies and loans for increasing the energy efficiency of the building stock and new houses in 2009 – in total more than 600,000 properties (Uttich, 2010).

Efficiency standards for household devices have been introduced by many governments, for example the US Energy Star program and the Japanese Top Runner Program, which stipulates that the efficiency of this top runner model should become the standard within a certain number of years (De Wachter, 2006). A relatively new policy instrument called White Certificates[2] also primarily targets electrical appliances by creating a policy instrument that rewards energy-efficient products via market mechanisms (Bertoldi and Rezessy, 2009).

One major market barrier preventing higher rates of renovation is the tenants' lack of willingness-to-pay. A possible way of overcoming the principal-agent constellation is by getting the state to subsidize the difference between the amount that tenants are willing to pay and actual costs:

> DEGEWO, a major housing association in Berlin, has held public discussions on the energy modernization of some large housing estates and the impact on rents. They estimated an increase of rents between 3 and 14 percent. This is the dilemma: surveys of tenants show that 80 percent of tenants want energy modernization, but only around 33 percent are ready to pay more rent in return. There are only two possible solutions: either the state steps into the breach and compensates for the difference in order to dampen the social impacts, or rents including heating remain strictly cost-neutral and the increase in the rents will be limited to 5 to 10 percent. (Prohl, GASAG)

Prohl suggests the same line of argumentation as the idea behind the McKinsey abatement-cost curves: If the overarching objective of political decision-makers were the cost-efficient reduction of greenhouse gases, they would more generously subsidize building efficiency programs:

> The core issue in Germany's energy transformation is who will bear the costs. A sensible and sustainable solution initially involves costs that will have to be borne by society in general. Up to now politicians have implemented the support for the switch to renewable energies fairly elegantly via a surcharge on the price of electricity. But the charge that has now reached 3.5 cents per kWh has made the wider public increasingly aware. Customers naturally do not want to pay high prices any longer. Care must be taken not to overshoot the target. The most efficient instruments should be promoted. Photovoltaic plants currently receive massive support, but they have little effect on the total electricity supply. If there is a desire to save energy, then one must rather start with efficient generation in order to achieve the highest reduction in CO_2 per Euro invested. (Prohl, GASAG)

Alexander Voigt of Younicos argues that the expansion of renewable energy supply may be faster than policy measures for building efficiency, though:

> In Germany, most houses belong to people, to the owners. They have to be convinced to invest. Can we modernize the total building stock in Berlin in

the next 20 years? No. In the last 20 years we have put up some new buildings and have renovated a few neighborhoods, cosmetically but by no means from an energy angle. If a house that is heated with gas or oil is converted to renewable electricity, I have achieved a lot for the CO_2 balance, although I may not have even touched the house's absolute energy consumption. That is more efficient and quicker to implement than the concept of modernizing buildings. But building efficiency should not be traded off against renewable energies. (Voigt, Younicos)

While much of the literature on the efficiency gap focuses on split incentives between landlords and tenants, little attention has been paid to the sector-specific industry structure. However, some idiosyncrasies of the buildings sector may substantially contribute to the persistence of the efficiency gap, according to Hengstenberg of building-efficiency service platform co2online:

> Similar to car manufacturers, pump manufacturers and companies specialized in fittings and heating technology spend lots of money competing with each other. For example, Viessmann spends millions of Euro on its representation at ISH, the world's leading trade fair for energy-efficient heating, air-conditioning technology, and renewable energies, which is never aimed at end customers but only at firms from the crafts sector. The manufacturers compete for orders from craftsmen for central-heating boilers by offering incentives and benefits, instead of supporting those firms to expand the market and increasing the number of buildings being restored in an energy-efficient way. Our criticism of this industry is that it is highly fragmented into the production of different components that have to fit together perfectly in the customers' homes. It does not see the whole chain of value creation. It rather focuses only on the subsequent stage: at the top, there are the manufacturers, then the wholesalers, then the craft firms, and the last in line are the homeowners. To a very great extent, the interesting ideas get lost on the way.
>
> Of the 300,000 crafts firms in Germany, only five to ten percent are actively looking for new business opportunities, the others are rather passive. There is no competition and little pressure to act. Neither is there any marketing, for example, in the sense that a craftsman actually approaches a homeowner and offers to make his house more energy efficient.
>
> Most craftsmen are fully booked for weeks in advance. They typically stick to tasks they are familiar with, and the ones they can earn quick money with. If a craft firm has fewer than five to seven workers, it only deals, so to say, with dripping taps. There are now so many different ways of revamping the total energy system of a house that the job is getting more and more complex. An average craftsman is already challenged when

installing a heat pump. In a one-man or two-man firm, the workers have a busy schedule all day long and no time to spend in front of their computers or to start a campaign in the local newspaper. That requires certain support. When I asked a craftsman from Frankfurt why he had not responded to my customer inquiry, he answered it was not possible for him to read his emails every day. (Hengstenberg, co2online)

Liberalization of some of these markets may create genuine action on the part of some firms to maintain or increase their customer base, like in the chimney cleaning market.

Even for smaller buildings, four agents are involved in the heating system: the firm that installs the heating, the one that services and maintains it, the chimney sweeper or cleaner, and the caretaker who switches the heating on and off. None of them knows what the other does, and normally they do not communicate with each other. For example, the chimney cleaner does not test the efficiency of the boiler.

The liberalization of the German chimney sweeping market in 2013 will animate chimney sweeps to develop new business models and service ideas out of fear that outside contractors will invade their territory. I see incredible potential for more turnover, more market and more energy efficiency precisely in this field – chimney sweeps, metering service providers, and maintenance services – since those technicians are in constant contact with their customers. This differs, for example, from the firm that installs the heating system and shows up only every 20 years or so. We cooperate with the chimney sweeps and devise new concepts together with them. (Hengstenberg, co2online)

Cooperations, alliances, and even outright mergers and acquisitions among crafts companies may accelerate learning and offer new services.

Some firms are joining up to form alliances and offer integrated products and services. Although they remain independent, they establish a brand, for example, and make it known in the end-customer market. This strategy not only includes simple installations, but also more complex energy-efficiency services, for example the maintenance and control of energy consumption.

For the sake of climate protection and energy efficiency, it would make sense to have a concentration process in the craft business and more firms with 50 to 100 employees. Only firms of a measurable size develop integrated solutions, take advantage of subsidy programs, and become active in communication and marketing.

Big German enterprises that are active abroad have recognized the market barriers of their three-level sales process. For them it is important to build up effective communication with end users. (Hengstenberg, co2online)

Creating lasting ties with the building owner (co2online)

To achieve longer-lasting returns in building-efficiency advice requires a business strategy beyond state subsidies. co2online is a non-profit organization that provides a web platform for homeowners and tenants to assess their individual energy consumption, compare it with prototypical indicators derived from the sample, and calculate which investments in building-efficiency pay off after how many years. The organization faced a difficult start and needed almost two decades to establish itself, according to its founder:

> Our business idea developed in 1992 at the time of the first Rio sustainability conference: if climate change really is happening – and we were already convinced of that back then – then the world needs a huge expansion of good advice. Our idea was to revolutionize conventional energy consulting, which even today is still organized as a personal service – much like Henry Ford II who switched from individual production to series assembly-line production for the Model T. The core idea was to use IT to process information contained in documents such as the heating bill but also in gas meter readings, the electricity bill, and other meter readings to increase transparency for the people concerned and ideally inspire appropriate action.
>
> Conventional energy consultants are not sufficiently unbiased to perform that task. No homeowner pays €100 or more to someone if it is clear that that person has a professional interest to detect deficiencies because that is how he makes his living. (Hengstenberg, co2online)

Hengstenberg explains how co2online came into existence.

> We were five failed academics and activists from the environmental scene and founded a civil code partnership, which we called the Working Group Energy. After some time I was the only one left from the group of original founders. In the first 10 years, we only made losses because the concept was new and required extensive explanations. We started with a dreary round of knocking on doors at the municipal authorities, trade associations, and housing associations. The breakthrough came when the German state-owned development bank KfW commissioned us to design an advisory website for building-efficiency. In 2003 we founded a non-profit company, co2online. Even though by then we had to win KfW contracts in official tenders, we were usually the only supplier of this service, which was our unique selling proposition, so to speak. For a big project from the city of

Munich, we converted the Working Group Energy into a limited company and called it Senercon. Today we have a total staff of around 60, a little over 20 with Senercon, and just under 40 with co2online. (Hengstenberg, co2online)

co2online's major asset is a database of energetic refurbishment measures in half a million buildings.

> In our communication with the final customer, we build upon economic rationality. We have exact knowledge of payback times of investments in energy-efficient buildings and cost-efficiency of refurbishment measures, empirically established from our database with a million buildings, of which we know the refurbishment history in 500,000 cases.
>
> Every year we advise about 1 percent of the German building stock. Cumulatively, we have now reached around 10 percent. We complete around 15,000 online advice sessions per week.
>
> With great reliability, we can assess the effect of individual measures for the type of building, the year of construction, and the age of the building. A complete building refurbishment has a payback time of around 15 years, while a new roof also takes a considerable time period, and it is much longer for windows. But for investments in heating technology, payback time can be extremely short. (Hengstenberg, co2online)

The experience with online advice services is unique in the European context, according to Hengstenberg.

> Several websites also offer online consulting services for building-efficiency. But they do not have our depth. In Europe, we are the only ones to offer such a comprehensive tool. Online advice systems often do not make calculations using the actual consumption data but rather estimated values, which of course makes them less precise. Our website offers users benchmarking with comparable buildings, based on a sample of one million buildings. In addition, our team updates the energy data and information on subsidy schemes, more than one hundred, many local or regional, on a daily basis. (Hengstenberg, co2online)

The digital age allows for instant, unbiased information on issues that concern homeowners and tenants.

> Most people use our advice tools, which they access from about 1,000 websites with a specific question in mind. They may have exorbitant heating costs or maybe they are moving house and want to have the new house assessed. Or someone has read about a new subsidy program in the

newspaper and wants information about it. Once users have entered our advisory system, they can choose various paths: either to get some brief information about support programs or to have the cost of the complete modernization of the building calculated. Ideally, in the end, users give us feedback. (Hengstenberg, co2online)

The organization positions itself as a first and immediate response to a customer's request and in no direct competition with any conventional player in the market.

co2online is a non-profit, limited liability company. Unlike a manufacturer or a dealer, we do not try to sell a technology straight away but provide neutral advice for the consumer, free of charge. We never act as competitor to conventional consultants but only deliver a first assessment, which the energy consultant can then build upon. For example, we suggest reading the natural-gas meter 10 to 20 times per year. Those records can later be used by the energy consultant to get a precise thermo-hydraulic fingerprint of the building – after we have related the metered fuel consumption to the medium outdoor temperatures of those intervals. Or people have a 15- to 20-minute session to store in our system everything they know about the building.

Our system could therefore be described as a first profiler. A proper energy consultant saves the time of a first site visit and can even submit initial suggestions by telephone. We see ourselves as a marketing agency for the crafts sector and for energy consultants. We prepare the market and aim to create demand on the customer side. We do not want people only to start thinking about a new heating system when the old one has broken down, but to give some thought beforehand to how to improve it and how much could be saved by a replacement, even if the system is still functioning. We draw the attention of the house owner to new products and to efficiency technologies. (Hengstenberg, co2online)

It efficiently occupies a market niche.

Compared to a conventional energy consultancy, we reduce costs by a ratio of a hundred to one. Our unit cost of one online advice session is around €1, whereas a comparable session with an energy consultant costs €100. Although the content is different and we only perform the preliminary stage in a consulting process, our service, for example, does not incur travel costs for the energy consultant's first assessment visit. A Danish study has found that energy consultants spend 70 percent of their working hours in their cars. (Hengstenberg, co2online)

All programs are neutrally assessed in their outreach and success.

> Our advisory services are anonymous. However, if our users wish to have the results of the evaluation sent to them as a data sheet, they give us their email address and allow us to get back to them for evaluations within six months to a year, depending on the kind of advice we provided. Around 20 to 30 percent of our online questionnaires are returned, while questionnaires sent out by post the rate of return is just under 15 percent. These evaluations assess, for example, whether users have adopted our suggestions and have converted their heating system to a heat pump with a solar thermal plant, or have followed the advice of the local craftsman and have installed an oil-fired boiler. According to our evaluations, we are responsible in about every 10th case of such an online session for the user deciding to adopt a modernization project suggested by us, requiring an investment of €10,000 to €20,000. (Hengstenberg, co2online)

Access to homeowners and tenants potentially interested in the organization's service is facilitated by its presence on various, relevant internet portals.

> Industrial enterprises, energy utilities and municipalities integrate our service into their websites. In return, they pay us an annual fee. Over the years, we have built up a network of 1,000 portal partners in Germany who integrate our tools. That gives us reach and relevance. (Hengstenberg, co2online)

The expansion into other European markets works most successfully if local partners cooperate, according to Hengstenberg:

> Thanks to an EU grant, we are now present in 10 European countries. But our advisory systems are only successful if they are not designed from a merely technical point of view, but are accompanied by communication and marketing. One has to be very familiar with the local situation, which takes time: What kind of communication is effective there? What kind of language does one need to reach the people? In comparison with Italy, Spain, and France, countries like Germany, Austria, England, and those in Scandinavia have a unique energy-efficiency culture. Yet our most successful operations are in Greece because our project partner there is the World Wildlife Fund, which as a partner is very strong at communications. The Bulgarians have now integrated a big telecommunications enterprise as a partner, 700 staff members are already registered there. Soon they will be contacting three million customers.

> The European Union badly needs efficiency data on a European level. Efficiency targets should be compatible throughout Europe: How has the renovation quota changed? What is the state of modernization? Where are the inhibiting factors? In what areas are high-consumption buildings located? Our data for Germany are an outstanding basis for that. (Hengstenberg, co2online)

The financing of co2online's activities still predominantly stems from public sources. The organization actively seeks out alternative financing models.

> From the outset, we were financially backed by the public sector, that is, the municipalities, the Federal Environmental Office, and later the Federal Government, which made it possible for us to come up with this energy consultancy. And that is how we are still financing ourselves today. We develop ideas and projects in-house and then seek out appropriate funding schemes.
>
> Nobody from the outside can imagine how to earn a single cent with a service like ours.
>
> In order to secure financing we also have to explore commercial business opportunities, since public funding amounts to a maximum of 70 or 80 percent only. For example, as sub-contractors we prepare data from our database and integrate new questions in our surveys for research projects. We have a stock of 150,000 email addresses of house owners and tenants at our disposal, which can be used to carry out surveys. For owners of these email addresses, we know the type of building, the ownership structure, and the heating system and the date of installation. To carry out academic or market research, the right sample can be targeted and made available. Of course, we do not sell any protected data. We rather assist in answering research questions.
>
> However, if a user asks what products are available in the market for a certain efficiency measure, and he explicitly states that he wishes to be informed about a product, then this opens a window to transfer him to companies specialized in that product or service. We inform the user that he or she will then leave the co2online website and will be passed on to an external provider. However, that service only generates minimal revenues and is, of course, not our original business purpose. But in principle, even a non-profit company can earn millions of Euro, as long as the surplus is reinvested consistent with the non-profit status. (Hengstenberg, co2online)

Apart from digital presence on companies' and utilities' websites, co2online also pursues other ways of generating revenue.

> We organize competitions in schools and other events, for example on climate protection. Here we receive support from companies in the form of sponsoring. (Hengstenberg, co2online)

A promising source of constant income seems to be to charge for repeated customer services.

> As a precaution for the risk of public grants drying up one day, we explore what additional advisory services our users would be prepared to pay for. Most of our advice tends to be of a one-off nature. They are done once and then perhaps repeated two or ten years later. Our goal is to regularly advise the user over a longer period. For this purpose we have developed an instrument we call interactive energy savings account. This tool with over 52,000 registered users by now – mostly homeowners and tenants and a few schools in 10 European countries – serves as a constant companion to each user's efforts to save energy. At the moment it is still free of charge but we are pondering to charge users an annual fee, if, in turn, for example, they receive free expert reports that entail dedicated consulting work by engineers and would hence have a certain market value. We may also offer a bundle of flat-rate energy consultancy services under the energy savings account: if a homeowner would like to cut electricity and heating energy consumption of his house by, say, 40 percent over the next three to five years with many small, intelligent measures – in fact, that is the process that most often takes place in reality – then over this period he would pay a user fee and be given expert support. When the refurbishment phase is completed, the success of the project is assessed by both our client and us. (Hengstenberg, co2online)

Hengstenberg comments that building efficiency has not yet found its place on the priority list of political decision-makers and on the public agenda:

> After 20 years, politicians have started to realize the value that we offer with our online tools and websites. But we have to explain it to them again and again. The attention of the general public is caught above all by new technological ventures and so-called lighthouse projects – but a constant drip wears away stone. (Hengstenberg, co2online)

Energy performance contracting

From engineering to economics

Energy performance contracting, also coined "efficiency contracting" or simply "contracting," typically contains an individualized contractual agreement

Box 6.1 **Building efficiency in developing countries – Druk White Lotus School in Ladakh, India**

Energy-efficient architecture of buildings does not have to be overly costly and can be successfully implemented in the rural context of developing countries, too. The Druk White Lotus School in Ladakh (India) won several prices, including the 2002 World Architecture Award for Best Green Building and Best Education Building. It was designed by the global engineering practice Ove Arup, which is also involved in high-end architecture like Beijing's National Olympic Stadium (the so-called bird's nest) and one of the world's tallest buildings, the 600-meter-high Canton Tower in China.

Ladakh, a remote region in the Indian Himalayas, has extreme weather conditions and has in the past been exposed to seismic activity and outright earthquakes. A team of engineers from Arup started in 1997 to design a complex of nursery and infant courtyard, a junior school, and a senior secondary school with dormitories for the pupils. The buildings have a southeast orientation to capture a maximum amount of sunlight – even during the cold winter months – and are equipped with dark, heat-absorbing materials to store the heat. A system of natural ventilation lets the air circulate in the rooms. Solar pumps are used to supply the school with the water it needs. Around half of the initial investment in solar energy installations was financed by a carbon-offset fund, and the school aims to further increase the installed capacity of photovoltaic panels and inverters. Except for emergencies, the diesel generator is switched off.

Figure 6.4 Celebration at Druk White Lotus School, India
Source: Druk White Lotus School (2012).

> Most importantly, the engineers tried to use materials that are locally available and, hence, have the least impact on the environment. Soil from the site was used in the roof construction. The mud bricks for the inner walls were hand-made in a nearby town. The granite blocks of the exterior wall are formed and finished from stone found on the site and gathered from the surrounding boulder field.
>
> Arup contributes to the success of the school by sending a senior design team member to visit at the beginning of each year's building season in April. This member is followed by another Arup employee who typically remains as a resident on site for around four months.
>
> (*Sources*: Arup, 2012; Druk White Lotus School, 2012; Indian Ministry of New and Renewable Energy, 2012)

between the owner of an existing building and a so-called contractor, who implements a retrofit of the building in order to reduce the operating costs and environmental impact of the building (Mayer, 2010, 2). Typically, those measures include "the replacement of boilers, insulation, cooling systems, and lighting and temperature automation controls, as well as the integration of energy data management" (ibid.). The contractor has the turn-key responsibility for the project, including technical tasks like preliminary energy audits, system design and engineering layout, installation of the devices, and performance measurement during operation, as well as the responsibility to set up a financially viable business case, including projections of energy prices, operation and maintenance costs, etc.

While the owner of the building benefits from multiple improvements in the energy system of its property and a decrease in its energy bill, as soon as the contractual relationship ends, the contractor is rewarded by the difference between projected and actual energy costs over an *ex-ante* specified amount of years. Figure 6.5 on the following page shows the principle features of energy performance contracting on a time axis.

The major challenge for the contractor is to tackle the inherent complexity and trade-offs between technically realizable tasks and overall long-term cost efficiency. For obvious reasons, numerous risks are involved and have to be borne. Most importantly, the contractor has to take over the performance risk of the endeavor, ensuring that the system allows for the projected energy savings. Further risks involve insolvency of one of the contracting parties and ownership changes. Project financing is most often organized by the contractor, but sometimes property owners or third parties like banks may get directly involved. Governments and municipalities may grant partial funding, fiscal incentives, and loans, too (Mayer, 2010).

The contractor has to find an optimal mixture among investment alternatives with largely differing capital costs, internal rate of returns, and technical challenges in the implementation, ranging from the installation of efficient light bulbs with a minimal payback period to cogeneration

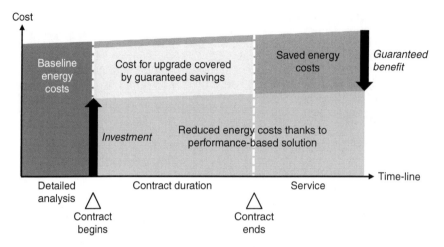

Figure 6.5 The concept of energy performance contracting
Source: Mayer/Johnson Controls (2010).

or micro-generation devices with a payback period between 5 and 25 years (Sweatman and Managan, 2010).

Incremental improvements of individual energy-using devices and design features – exploiting the cheapest and easiest technologies – are often called "cherry picking." Although they may hamper the business model of the contractor, from a carbon-emission savings perspective, they are a fully legitimate means to decrease the environmental impact of a property and allow for savings without entering into an outright contractual, long-term relationship with an outside organization.

Energy companies like E.ON start perceiving themselves as providers of efficiency and energy performance contracting services:

> Our sales departments provide a bundle of products and services for specific customer groups, including for example the smart home. Since customer requirements differ widely, we offer different products. A medium-sized bakery requires different products compared to a small mechanical engineering firm or the owner of a condominium. In contrast, our large customers with own energy departments tend to prefer products tied to the wholesale market. For quite some time, we have been offering our industrial customers cogeneration and energy service contracting models. (Rümmler, E.ON)

The German energy agency estimates that in the public sector, including schools, administration buildings, prisons etc., 85 percent of the economically attractive properties were not yet under a contracting regime in 2007 (Dena,

2007). Consulting practice Pike Research estimates that if all commercial space built until 2010 in the United States had been included in a 10-year retrofit program, savings in energy expenses would have had the potential to reach more than US-$41.1 billion each year (Pike Research, 2010a).

The Institute for Building Efficiency – through an initiative of US company Johnson Controls, a provider of equipment, controls, and services for heating, ventilating, air-conditioning, refrigeration, and security systems for buildings – names four major obstacles that have contributed to the lack of success of energy performance contracting in Europe (Mayer, 2010, 2).

Except for Germany, most public and private actors in Europe do not realize that energy performance contracting is possible and that companies offer this service; hence a lack of awareness prevents markets from developing. In addition, policies and fiscal incentives targeted at promoting performance contracting are often missing from the policy agenda; public and private property owners do not have the capacity to find a proper contractor; and regulation is fragmented, with a lack of common definitions and harmonized processes across or within EU countries.

Compared to Europe, energy performance contracting of institutional buildings, especially federal buildings, in the United States received a financial stimulus with the American Recovery and Reinvestment Act.

Konrad Jerusalem, founding member of Kofler Energies and CEO of energy performance contracting firm Argentus, predicts a stepwise expansion of efficiency services, from industrial process optimization to commercial applications. However, the implementation of residential efficiency measures will not become profitable – unless energy prices rise substantially, or prices for the relevant technologies decrease:

> If the aim is to save energy by technical optimization, one has to start at the point where it is relevant for the customer. Already 20 years ago, energy-intensive industries like the paper industry and aluminum smelters realized that energy costs are a substantial part of unit costs. So these industries recruited the best engineers to optimize production technology. The next stage of technical optimization could tackle hotels, hospitals, etc. For these customers, their concern is not an improvement of production technology but essentially the in-house use of energy, for example providing heat for buildings efficiently or cutting energy consumption for ventilation. In this market segment, there is still great potential for optimization. The private customer would be next on the list. However, I believe a lucrative business with private customers will still take a very long time because energy costs of a private customer are a relatively small portion of his living costs. Either energy costs will have to soar or technologies will have to become far less

expensive, if this market segment is to become cost-effective. (Jerusalem, Argentus)

Jerusalem analyzes why efficiency contracting has not yet found broad appeal in the German conservation market:

> The technical optimization and reduction in energy consumption of existing plants, machinery, and technical devices has long existed in the so-called contracting business. Yet it has never really worked on a broad basis in Germany. We have analyzed why this was the case and have looked for a new approach.
>
> Energy performance contracting is still a very technology-oriented, complex business. There is a relatively high degree of individualization and hardly any standardization. Every project is different. This may be due to the fact that the business model is very much technology-driven. In most cases, highly qualified technical specialists try to find the best technical solution – which does not necessarily yield the maximum benefit for the customer or result in the most economical solution. (Jerusalem, Argentus)

Energy performance contracting requires trust from investors and customers because both parties have to be willing and confident to enter into a long-term business relationship:

> In the case of contracting, one has to enter close ties with a customer for a relatively long period of time. In order to get accepted as a project partner, a contracting provider has to convey a lot of trust since the customers must be convinced that they can hand over something very important, namely security of supply, to a service provider. This is a relatively high hurdle, since contracting is not without risks from the technical angle. For example, a combined heating and power plant has to be integrated into the customer's energy system. In Germany, that is emotionally highly charged. There is a fear that the electricity will run out or the heating will no longer work. But in this country, nobody is left without electricity because if a customer leaves the contract with his supplier, the local utility is obliged to deliver the electricity. It may be a bit more expensive, though.
>
> If long-term contracts are concluded, each business partner has to trust that the counterpart will remain solvent for 10 years or more. It is extremely difficult to make a forecast for such a period of time, even in areas that are relatively stable, for example hospitals or hotels in prime locations, which are likely to continue to exist irrespective of the operator. But even there, the market of customers with good credit ratings is relatively limited.

Similarly to banking, in contracting one has to ensure that there are collateral securities. According to German law, when a moveable property such as an electricity generating device is firmly based on a plot of land, then the owner of the land automatically becomes the owner of the power station. Fortunately, there are some exceptions to this legal principle that require third-party consent. But if the deal depends on the consent of third parties, for example in hospitals run by the church or a municipality where big advisory boards and politics influence the decision, then obtaining agreement can often be very cumbersome. (Jerusalem, Argentus)

The complexity of the contractual and financial obligations resembles large projects, even though the scope of the projects is much smaller:

In my early career, I structured major power plant projects to make them bankable as part of a project's financing. It is a very complex operation to structure projects in such a way that all the risks that have been identified are then mitigated. Normally, this is only profitable from a certain size upwards, for example with a power station with a capacity of several hundred megawatts. However, energy service contracting deals with very small power plants, often 0.5 to 1 MW, and rarely more than 5 MW. In the financing of these small plants, one encounters a similar degree of complexity, but financially there are not the same possibilities for maneuver. (Jerusalem, Argentus)

Jerusalem confirms the findings of the US National Association of Energy Services Companies that institutional and psychological barriers have to be overcome:

As a rule, when the customer is, for example, a hospital, a big hotel, or an industrial firm, then one deals with a technical director whose job over the last 10 or 20 years included optimizing energy consumption. It has proven to be psychologically challenging if an external service provider arrives and tells that person what he has to improve. It is fundamental to get that person on board and convince him. (Jerusalem, Argentus)

In the future, Jerusalem expects the energy efficiency business to become a service that will be outsourced by many companies and utilities, similar to services in information and communication technologies:

Many customers are still hesitant to leave the implementation of energy efficiency measures to an external service company. One can argue that many companies have outsourced, for example, their IT services. However, we know that in the case of IT, it took a long time to reach the status of IT outsourcing we have today. In the case of energy, we have not yet reached the stage where outsourcing is easily accepted. (Jerusalem, Argentus)

Andreas Prohl of Berlin's natural gas supplier GASAG points to obstacles of efficiency contracting in the legal framework:

> From a legal point of view, energy service contracting in the case of rental properties is difficult. Normally, the owner has a heating facility based on oil or gas combustion in the basement. The tenant bears the investment cost of the boiler through the rent and pays for the fluctuating consumption of oil or gas through extra charges. Contracting pursues the idea of calculating a heating price for the landlord that includes not only the cost of the fuel but also the share of the capital expenditures and operating costs. But the landlord cannot simply pass these expenses on to the tenant. The situation with tenancy law is very complicated on this point, and in Germany, for example, it depends on whether the tenancy agreement was concluded in what was West Germany or what was formerly East Germany. There is no uniform basis.
>
> We aim at leaving the rent, including heating, unaffected. If we have higher capital costs for a new heating system, this increase is offset by lower energy consumption and thus reduced fuel costs. Modern boilers can save so much that they can compensate for the capital costs for the new facility. However, that is the ideal scenario – in many cases one that is not so easy to put into practice. (Prohl, GASAG)

According to Jerusalem, the energy-intensive industry could be incentivized by a scheme that promotes energy efficiency via a reduced tax component in the electricity prices:

> It appears that the efficiency industry needs some external assistance in order to gain momentum. For example, a certain relief could be provided on levies and taxes, etc.
>
> A large fraction of around 40 or 50 percent of the price of electricity is composed of taxes and levies. For example, the levy imposed under the Renewable Energies Act (EEG) is 3.5 cents per kWh and the electricity tax is 2 cents per kWh. Including value added tax, that already amounts to a share of 9.4 cents in the 24 cents, which is the average price paid by residential customers for electricity. Commercial customers pay lower electricity prices, so that the share of tax for them is proportionately even higher. Here I see an opportunity to offer an incentive for efficiency investments, for example by imposing no EEG levy or a reduced rate of electricity taxes.
>
> If a power plant is directly connected to a local load, transmission charges and the renewable feed-in levy can be avoided. That is already an attractive incentive. In addition, loans with favorable conditions could equally support service providers and thus promote the market for efficiency. (Jerusalem, Argentus)

Municipal utilities are perceived as serious competitors for other new entrants, in particular independent efficiency contracting companies like Argentus and Kofler Energies. Jerusalem hints that utilities have multiple objectives in their strategies within the efficiency contracting business.

> There is stiff competition from municipal utilities. They have access to favorable loans and do not have any requirement to yield high returns from their operations. Big utilities have clearly defined minimum hurdle rates, which impose that every project must earn a certain return on capital. That is not at all the case with local utilities, particularly not with the smaller municipal utilities. They are far more concerned with keeping their local customer base. Many decisions they take are based on political considerations. An important customer will always be kept, irrespective of the cost. To compete with such a municipal utility is difficult, if not impossible. (Jerusalem, Argentus)

However, the traditional business of utilities is to sell energy. Jerusalem doubts whether municipal utilities can reorient their strategy and become efficient service providers:

> The classic electricity supply industry aims to sell its product at highest prices possible. That is what their business model is about. The goal of the efficiency industry is the opposite: saving energy and lowering prices. Those two strategies obviously cannot go hand in hand. (Jerusalem, Argentus)

Apart from the municipal utilities, a number of other players also shape the competitive context of efficiency contracting, especially small and medium enterprises from the manufacturing business.

> There are many small firms in the energy service contracting field that build power plants or are somewhat involved in their construction. However, their creditworthiness often tends to be shaky and some of them actually do go bankrupt. (Jerusalem, Argentus)

Due to their profound knowledge of local players and administration, municipal utilities face ideal conditions to enter the efficiency contracting market. The municipal utility of the medium-sized town Krefeld has started seizing these opportunities. The utility closely cooperates with local technicians to offer its services:

> We are working on concepts similar to LichtBlick with our partners. These are fuel cell projects, calorific gas boilers, etc. We believe that these have

> potential, but we need local partners for proper implementation. Via these local partners, for example, plumbing and heating firms, we want to offer micro CHP units also to private households in the future, using contracting or other innovative solutions. That could be fuel cell solutions, a WhisperGen installation, or other solutions that have already been tested. We are clear that the business will only develop slowly and a shift will only be possible if there is a decision to invest, for example, in a new heating installation. (Liedtke, SWK)

Compared to other countries, Germany benefits from the fragmented ownership structure of the distribution network, which allows new players to occupy niches:

> The United Kingdom is fairly advanced with respect to energy efficiency services. By contrast, France still faces a monopolistic situation because of its state-owned utility EDF. Smaller service providers emerge slowly. Germany is the country of small and medium-sized enterprises. I have great confidence in Germany, precisely because we have a lot of municipal utilities that are already highly fragmented, making it easier for new enterprises to establish themselves. Ultimately, business flourishes where money can be made. Since efficiency measures are not subsidized – in contrast to wind and solar power generation – not much money can be made at present in this field. We lag behind the efficiency targets that have been set. Something needs to happen. (Jerusalem, Argentus)

Indeed, in a number of new initiatives, the German government strengthens the options of private homeowners, notably via credits and reduced loans or direct state aid for the installation of new devices in the heating system. For example, if a building is retrofitted to reduce energy consumption by 45 percent, the German public bank KfW subsidizes 20 percent of the total costs (Hofmann, 2012).

Product standardization as the key to customer management (Argentus)

References are an essential prerequisite for an energy performance contracting company in order to gain market access.

> In our consultancy practice we often see that one of the main difficulties of start-up contractors is the lack of strong references. The client wants to see convincing reference projects before entering into a long-term contract.
>
> In addition, especially large customers are looking for a services provider that can take care of a wide range of services: energy efficiency, technical

consulting, energy supply at low tariffs, that is, the "one stop shop." (Jerusalem, Argentus)

Lean management and an efficient handling of client visits is one of the key components of Jerusalem's strategy.

> Internal efficiency is also hugely important for a profitable efficiency business. Lean structures have to be created in the firm itself so that one does not have to make five trips to each customer but just one or two.
>
> Before Jerusalem entered the market for efficiency services, he analyzed in which part of the value chain new business opportunities could loom. "When one intends to start a business in the efficiency industry, a viable and profitable business plan has to be developed. It is sensible to focus on corporate customers first, for example hotels and hospitals, because usually these have sound credit ratings." (Jerusalem, Argentus)

A professional marketing of the service provided seems to largely fall outside the competences of entrepreneurs who have established companies in the field.

> The whole contracting and efficiency market is still a rather boring and very technically oriented business. It offers sound technical solutions but is not as attractive as high-involvement products such as smart phones, where the customer says: "That is really fun!" One key success factor for a strategy in the efficiency business is to approach potential customers with professional marketing and not just with a technical solution. The customers do appreciate that approach, since it is a relatively rare phenomenon in the efficiency market. (Jerusalem, Argentus)

Jerusalem applies a strict economic rationale to efficiency service contracting, following the idea that changes in 20 percent of the building equipment lead to 80 percent of the conservation target.

> A key factor is to rigorously standardize the services a company offers. One should focus on a few but effective efficiency measures and not invent something absolutely new for every customer. The so-called 20–80 rule should be applied, that is, with 20 percent of the measures 80 percent of the effect can be achieved. Further measures tend to become uneconomical. Customers will understand and support that approach.
>
> First of all, it is necessary to create transparency, for example by installing smart meters so that customers can get a precise overview of their four or five most energy-consuming loads. They can then alter their consumption patterns accordingly. The service provider only has

to install meters and create transparency – the rest is taken care of by the customer himself. That first measure already can save between 5 and 10 percent of the energy.

At the next stage, the service provider carries out minor investment measures, for example by optimizing ventilation with vents or improving energy flows in the heating system. Although heating systems have an optimal temperature difference between incoming and outflowing water, most systems are not adjusted to this optimum. These measures cost only the time spent on them and do not require any great investment. In this way, further savings of between 5 and 10 percent can be achieved.

Then an offer is made to the customer to actively optimize their energy management by themselves. Various forms of energy consumption are monitored. Then the system is either adjusted accordingly by the service provider, or the customer corrects it directly. This can take place without any inconvenient interruptions of the heating or cooling system of the premises.

Only after all these measures have been implemented – frequently not until after half a year of cooperation – a phase with somewhat greater investment can be initiated. But instead of replacing the whole system, one can start with minor investments. For example, pumps in the heating system are often old and consume a lot of electricity. If they are replaced, the amortization period is relatively short.

Only when all measures have been completed may the contractor suggest installing a cogeneration plant or a new air-conditioning system. With these measures, economically viable limits of efficiency investments are reached. (Jerusalem, Argentus)

Value creation in energy efficiency contracting can be profitable only if the investor reaps financial benefits from all elements in the value chain, according to Jerusalem.

As an energy service contractor, only a holistic approach yields appropriate financial returns. Value creation starts with construction work, where a contractor should already be involved. As the operator of the cogeneration device, the contractor provides heat or steam or other things the customer requires, but can already get involved in the delivery of the gas, too. If structured intelligently, the contractor can earn a bit in all sections of the value chain. Without this combination, the business model is not really economically interesting. (Jerusalem, Argentus)

Risk reduction via guaranteed energy savings (Argentus)

Many conflicts between the provider and the customer of efficiency contracting services emerge because of incomplete contracts and failed targets. Konrad Jerusalem explains how standardization of contracts can help in avoiding lengthy ex-post negotiations.

> It is beneficial to standardize products and keep things simple. Concentrating solely on savings has been practiced for a long time. But it can be complicated, since it frequently leads to frictions due to opposed or diverging interests. Hence, one should implement a standardization aimed at guaranteeing the customer savings at a fixed percentage, for example 10 percent. The engineers should calculate the maximum energy savings that can be extracted and investments required to achieve it. Then the contractor can tell the customer that at the start, his energy bill is, say, 100. Then 10 is subtracted from that amount and it is guaranteed to the customer that he will have to pay only 90 in the future, while the contractor promises to take care of all the rest. That is not so transparent for the customer, but many customers are still happy to hand over the job to somebody else and not to have to bother about it. They know that their costs are fixed. This approach incentivizes the contractor to increase overall efficiency beyond that 10 percent to generate a margin for himself.
>
> One of the critical success factors of a decentralized energy efficiency industry is consistent customer orientation. Communication is at the heart of energy service contracting. What counts is not always finding the solution that is technically the best, but listening to the customer and being guided by his needs. Since this field still yields low returns, it is favorable to create lean internal organizational structures. This is one of the reasons why I do not anticipate the big utilities entering this market segment. Their employees are way too expensive. (Jerusalem, Argentus)

In order to avoid a lock-in situation with its equipment, the company did not favor one technological solution. ICT services seem to be a promising field for the future.

> I have always been opposed to backing one particular technology, since technologies are evolving at rapid pace. The focus should rather be on measuring and controlling. Smart meters, for example, are experiencing declining prices and will perhaps in the future become a commodity used by everyone to measure energy consumption. At the moment, we cannot yet imagine how this is done, for example with wireless communication,

> **Box 6.2 Weather forecasts to increase the energy efficiency of intelligent buildings**
>
> "Better to optimize than to invest" is the slogan of MeteoViva, a German start-up that was founded in 2001 and has seven employees. Instead of capital-intensive investments like insulation, the company offers intelligent energy conservation with the help of meteorological forecasts.
>
> The engineers develop an individualized mathematical model for each building and its respective physical properties and energy characteristics, and use local weather forecasts to calculate the exact amount of energy needed to create the desired room temperatures. They take the buffer and storage capabilities of the building into account, as well as the heat generated by humans, lamps, and computers, and the warming effect of direct sunshine via the building's windows. With an intelligent use of each building's control systems for already existing heating installations, temperatures are optimized during the core working hours between 8 a.m. and 5 p.m. to create a comfortable working environment for employees.
>
> The track record of MeteoViva's achievements includes the municipal tax authority of the German city of Aachen. The building has 16,000 square meters of floor space with approximately 1,000 employees. When it opened in 2006, it already had all modern amenities for energy conservation like concrete core activation and passive cooling. But after MeteoViva had installed a weather forecast control, cost of heating could be slashed by another 18 percent. The amortization period of the investment was as short as two years – even after MeteoViva charged a success fee equivalent to a third of the savings in energy costs. Its control devices for a hall for high-speed train maintenance in Krefeld, equipped with 79 sensors, reduced the building's gas consumption by 47 percent, while electricity for air heating decreased by close to 90 percent.
>
> The company is also involved in the energy design of prestigious buildings like the new headquarters of the European Central Bank in Frankfurt and the permanent "BMW World" exhibition space in Munich.
>
> The market potential for energy savings solely in commercial and industrial buildings in Germany could be as high as €1.2 billion annually in avoided energy costs, if heat output were cut back by only 10 percent. In addition, savings in residential housing could be as high as €3.5 billion per year, also assuming energy savings of 10 percent, according to Markus Werner, the CEO of MeteoViva.
>
> (*Sources*: Hartmann, 2012, MeteoViva, 2012)

but that is precisely why one should not cling to one special technology but should always keep up with the state-of-the-art technology. Moreover, one should think in holistic terms. The cooling system should not counteract the optimization of the heating. That is inefficient. The best way is to take all components into account and then talk with the client.

Some players increasingly distance themselves from the asset business, where one has to invest and build, and move toward business lines that focus on operational management. (Jerusalem, Argentus)

The vision of Jerusalem is that energy efficiency services will reach a status that information technologies and the related services have already reached.

> Services such as IT will at some time in the future be handed over to third parties. IT in particular is becoming ever more important in our sector. Many firms decide to outsource it to professionals. This requires mutual trust between partners and very good agreements. At some point, technical directors will hand over the day-to-day business to external service providers. Up to now that has hardly been realized because of a lack of standardization, as opposed to, for example, information technology. There will be an increasing number of service providers who optimize the systems for their customers and take complete care of their efficiency. In other words, they will offer all-inclusive, no-worries packages. (Jerusalem, Argentus)

Findings on building efficiency and energy performance contracting

- *Measures to increase building efficiency face significant market hurdles*: Building efficiency will continue to face major market barriers because of dispersed ownership of the housing stock, split incentives, and the horizontally and vertically fragmented structure of the buildings industry. Apart from command-and-control regulation, information becomes the most important driver for the implementation of energy savings. Adequate web platforms for transparency can provide unbiased information for homeowners and tenants about their energy choices. To guarantee neutrality, the state should direct its efforts to enhance and broaden these services, especially for the new generation of digital natives.
- *Low opportunity costs of energy efficiency should make it policy priority no. 1*: If lower carbon emissions are the primary objective of the future energy system, energy efficiency should be prioritized on the policy agenda because it contains the greatest savings potential for the lowest costs. If major capital investments are directed toward supply installations that will become obsolete under more restrictive efficiency standards later, the sequencing of regulatory incentives should be reversed in order to avoid a costly lock-in effect.
- *The "Low-Hanging Fruits" offer an immediate potential for cost-savings without major investments*: While in the past, energy performance contracting was dominated by engineering principles – aiming for technical perfection and trying to cope with a high degree of complexity – a stepwise approach following the rule that 20 percent of measures already yield 80 percent of efficiency savings seems like a viable method to pull the contracting business out of its niche status. Marketing and a strictly economic rationale

will strive to "satisfice" rather than optimize. Behavioral changes, lighting, as well as the optimization of heating, ventilating, and air-conditioning do not require significant upfront investment and will pay off for tenants, building owners and operators, and contractors.
- *The standardization of energy performance contracts creates financial incentives for the contractor and hedging opportunities for the property owner*: If a target of 10 percent energy savings is contractually fixed by the contractor, say, the property owner benefits through increased planning certainty, while the contractor is encouraged to reduce energy consumption in order to reap the additional benefits resulting from excess savings.

7
Insights from Germany for a Decentralized Energy Future

Germany intends to radically transform its energy system over the next decades. The targets are clear – renewable energy sources are supposed to provide the bulk of generation capacity and electricity output. Any fundamental change of a large technical system requires major investments in infrastructure, though. How can such a transition be managed without overstretching the wallets of German energy consumers? Which business models and strategies emancipate themselves from the "sweet poison" of subsidies and succeed intelligently in the market? What will be the role of government and regulation in the new energy system?

We asked key stakeholders, entrepreneurs, and managers who have become engaged in working toward a decentralized energy supply about their views on how the system will evolve. While each of the previous chapters had a technology focus, we have aggregated the findings of our interviews and would like to point out 10 insights from Germany for a decentralized energy future – insights that we believe are helpful in understanding the dynamics of decentralized energy generation (numbers 1 and 2), business strategies to harness the market and profit potential of decentralized energy generation (numbers 3 to 8), and implications for public policy and regulation (numbers 9 and 10).

1 The decentralization snowball

We do not face a one bang revolution, but a revolution of a million stings

Multiple agents contribute – often in a dispersed, uncoordinated, and value-driven manner – toward starting a snowball effect that will overhaul the existing supply structure and permanently alter the competitive landscape in the energy system.

Start-ups and entrepreneurs like LichtBlick and Younicos create genuinely new business models that are adapted to liberalized electricity markets. Among the 800 German municipal utilities, many discover local embeddedness as their unique value proposition. More than 90 German bioenergy villages are engaging in the process of establishing largely autonomous island systems. Diversifiers from other industries, like Volkswagen, are lifting the manufacturing of decentralized technologies up to mass production levels. Equipment producers and service providers like Siemens are benefiting from their diversified portfolios along the energy value chain and – given their culture of innovation – are aiming to become leaders in research, development, and marketing of an integrated energy world for households. Car manufacturers like Daimler are close to the market launch and mass series production of electric vehicles.

Even incumbent energy utilities sense that they will be left behind if they do not join the movement. German energy incumbent E.ON is establishing a new business unit that is responsible for the selection of distributed energy initiatives to be commercially rolled out across its markets. Regulated grid operator EnBW Regional is pursuing a late-leader/early-follower strategy. But it has no other choice than to deal with the consequences of a decentralized energy supply, because private consumers in its service area are becoming producers of energy with photovoltaic panels on their barns and locally financed wind rotors on the nearby hills, which urgently requires that the company invests in an intelligent network.

The momentum for decentralization is gaining pace across the whole spectrum of corporate decision-makers and businesses. It nevertheless hinges upon the idealism of interested individuals inside and outside corporations and communities to convince their peers to reclaim sovereignty over their energy supply.

The decentralization snowball will inevitably shift the balance between central and distributed energy generation – if aided by subsidies, the avalanche will come sooner; if it relies solely on intelligent business models, it may gain momentum later, but as Frank Hose of ODR comments, "Decentralization cannot be stopped."

The interviews reveal that decentralization is indeed a disruptive innovation, a technical leap forward based on sophisticated system architecture with high-tech equipment, industrialized manufacturing processes, and multiple communication channels at its core.

2 Emotionalization of energy

Decentralized generation emotionalizes consumers by giving them the opportunity to become agents of change and to contribute to a better living environment

Energy moves away from an abstract flow of invisible particles to an issue of personalized identification. As much as locally harvested food is more satisfying for consumers than apples and strawberries produced in a different hemisphere and transported across oceans, locally produced electricity and heat offer the notion of self-determination and ecological consciousness. They are deeply enrooted in the overarching quest for empowerment, which started with skepticism against untamed globalization, was reinforced by the possibilities of the digital society, and has now become a mass phenomenon.

Consumers turn into producers of energy. They engage in activities that may not always be financially beneficial, but that have an emotional return. Motivated agents of change encourage entire villages to abandon the convenience of receiving energy from an unknown origin, to finance local industry and craftsmen, and to rely on renewable supply. They stimulate participatory processes in communities and put the origin of energy back onto the ethical agenda of responsibility and citizenship.

After often having been outperformed by the financial strength and economies of scale of big energy utilities, municipal utilities and local administrations are now benefiting from the advantages of local knowledge and proximity to their customers – they are discovering the marketing potential of an individualized energy world for their respective residents and thus can differentiate their value proposition from their competitors.

3 From single technologies to systemic viability

Decentralized energy yields positive system externalities

The whole is more than the sum of its parts.[1] Some of the devices used in a decentralized energy supply are not yet commercially viable on a stand-alone basis. But an isolated cost–benefit analysis fails to account for systemic advantages they entail. For example, smart meters and the respective communication infrastructure can be used for peak-shaving, that is, lowering temporary demand or supply spikes. Retailers and utilities can thus avoid costly electricity purchases on the spot market. Smart meters are part of an integrated strategy to offer a green and convenient energy world to customers, which creates value beyond a differentiated tariff for electricity consumption. Under similar cost comparisons for grid infrastructure, storage systems can prevent expensive construction and reinforcement of transmission lines.

Utilities and energy incumbents are only slowly starting to realize systemic benefits of decentralized energy supply. While their core expertise – central energy generation – was based on economies of scale and the optimization of the large-scale power plant portfolio while ensuring efficient transportation and distribution, decentralized energy generation poses an unprecedented challenge to integrate atomized energy producers into a complex, meshed network. System viability will count, and not technical and economic viability of a single technology. If positive network externalities of decentralized energy devices are recognized, utilities can play an important role in the diffusion of products and services that stabilize the grid – and accelerate technological progress by fostering mass production in the manufacturing of new devices.

The value proposition of decentralized energy moves beyond maximizing the sales of kilowatt hours to final customers. Future revenues will shift from a commoditized product – energy – to bundling geographically differentiated customer requirements into an integrated, individualized service.

4 Exploiting market volatility

Spot markets and secondary electricity markets offer financing opportunities for virtual power plants and storage equipment

Some decentralized applications are already economically viable on a stand-alone basis if intelligent market strategies are applied. Residential combined heat and power plants – micro CHP units – are the primary example among the technologies under consideration. The patchy heat or cooling demand of a single household would not justify the necessary operating hours to economically optimize the energy output. Especially established market players have doubted that micro CHP units could break out of their niche position without substantial subsidies. Yet, new entrant LichtBlick has successfully developed an alternative strategy and ensures profitability in settings where traditional business models have failed.

Their idea is to create a virtual power plant out of a cluster of micro CHP units, so-called swarm electricity, and use peak electricity prices on the wholesale market to finance the units. The business model has already found a number of imitators, even among traditional energy incumbents like Vattenfall and diversifiers from other industries such as Deutsche Telekom – an indication that the strategy is likely to pay off. LichtBlick has installed 420 micro CHP units and plans to reach a rollout of 100,000 units – which would exceed the annual, worldwide sales of micro CHP units by a factor of four.

Similarly, local electricity storage in integrated systems is currently not commercially viable because of high battery costs, unless companies use innovative ideas to finance their investments. Energy company Younicos, whose slogan is "Let the fossils rest in peace," has a long-term strategy to become one of the major players in an electricity system that predominantly relies on fluctuating renewable energies – and correspondingly requires adequate storage solutions. The company will gain experience with the complex handling of storage by entering "secondary" electricity markets, in particular the so-called balancing market, which provides backup power for unforeseen system imbalances due to a mismatch between supply and demand. The revenues generated by participating in the balancing market are then sufficient to finance its stationary storage devices.

Thus, liberalized electricity markets already today offer opportunities for entrepreneurs beyond mere subsidy schemes to experiment with innovative business strategies and succeed with their visions.

5 Guidance in the information tsunami

Expertise in data processing will ensure value creation in the future energy grid

The smart grid will unleash a "data tsunami" for utilities and consumers. It will add a second layer on the existing grid that consists of massive information flows with multiple agents – the prosumer, the grid operator, the retailer – interacting, coordinating, and exchanging data.

Especially for smaller utilities, this new information and communication layer poses an unprecedented challenge. Companies like smart meter producer Itron will provide a bundle of new services to utilities – ranging from initial consulting services about what meters to install to managing data flows, reintegrating them into the existing information systems, and providing a user interface for the final customer.

Larger utilities can create new income streams by providing these services and offering integrated data handling to less advanced peers and companies active in demand-side management.

Especially in the transition phase, with unclear standards, not yet fully mature technologies and uncertainty about the future regulatory framework, companies with knowledge about the functioning of the existing IT infrastructure and organizational practices of utilities will have multiple opportunities to generate additional income streams.

Integrated metering services, that is, power plus gas plus water, and differentiated billing can be a longer-term business prospect for diversifiers like Siemens and new entrants from the telecommunications sector. They will develop integrated solutions that ease the final consumer's burden in dealing with an increasingly complicated metering system.

Data protection is a perceived threat to constant data flows. Of all players in the field, municipal utilities can build upon their local roots and see their competitive advantage in conveying the corporate image of competence, continuity, and trustworthiness to their customers. They can offset their losses in revenues from grid operations and sales by gaining the reputation of being a reliable partner to intelligently manage exponentially growing data transfers.

6 From optimizing to satisficing

All elements of successful energy performance contracting already exist – they just have to be recombined in a commercially sound package

Some solutions for decentralized energy services face substantial market barriers due to their inherent complexity. Energy performance contracting is the prime example of a business idea that was developed more than 20 years ago and – despite its apparent appeal in terms of energy savings and cost reductions for the final consumer – never came close to its actual market potential. Entrepreneurs like Kofler Energies and Argentus have successfully reduced the complexity trap in contractual relations of energy performance improvements by rigorously standardizing the procedures. They distance themselves from lengthy and costly optimization procedures under engineering principles and apply a stepwise, least-cost approach that picks, so to speak, the low-hanging fruits first.

Satisficing means that 20 percent of efficiency measures can achieve 80 percent of the savings. Further improvements tend to become uneconomical. If energy performance contracting is rigorously standardized, its appeal to customers, especially in the public housing sector, will increase.

With a proper incentive scheme, risks can be reduced and further savings potentials realized. New players in energy performance contracting might guarantee a secured reduction of energy consumption of, say, 10 percent for their customers. This strategy creates an incentive for the contractor to exceed the agreed target levels, while providing a hedge for their clients to achieve guaranteed cost-savings and, over the long term, to benefit from a further decrease in energy consumption.

Many customers are still hesitant to leave the implementation of energy efficiency measures to an external service company. But similar to ICT services, specialized players will increasingly take over those tasks once the standardization of energy performance services will have increased.

7 Innovation and dissemination networks

Innovation and dissemination networks create substantial learning effects, reduce investment and financing risks for all parties, and accelerate the diffusion of technologies that are not yet technologically mature

If technologies are not yet sufficiently mature and suffer from uncertain market conditions, cooperations between manufacturers, retailers, and utilities can reduce risk for all parties involved. For example, Greenvironment sells and operates micro turbines produced by manufacturer Capstone. In a niche market, Greenvironment provides financial security for the producer to allow for further product development and innovation, while Capstone can adapt its turbines to the requirements of the retailer. By operating the turbines, Greenvironment gains valuable insights into the functioning of the devices, which it can in turn report back to the manufacturer. Cooperations of manufacturers with utilities also lead to greater chances of survival for niche devices in the market.

Joint ventures and alliances among manufacturers and between manufacturers, retailers, and utilities may also overcome market deficiencies. While not directly entering the manufacturing of batteries, Daimler, for example, has clear requirements for the configuration of an electric vehicle's battery, and it is taking part in the development process via cooperations with battery producers.

The internet offers new channels to create innovation networks. For example, Siemens promotes the concept of open innovation and organizes so-called innovation jams, where external stakeholders communicate their opinions and ideas on internet platforms. The company then integrates these ideas into its business and develops them further.

Dissemination networks allow companies to learn from early market experiences. Cooperations between utilities and operators of new devices like micro turbines are win-win solutions because they offer low-carbon energy supply for utilities and increase the expertise with the new technologies. Bioenergy village Jühnde has created a so-called marketplace for around 3,000 visitors each year, where it communicates its strategy, shares its experience with other interested villages, and promotes local businesses seeking opportunities abroad.

8 The right sequencing of the energy transformation

To avoid sunk costs in excessive transmission infrastructure, regulation should promote island systems, prosumers, and efficiency

In the German context, the energy transformation foresees the decommissioning of all nuclear reactors and a massive extension of offshore wind power. To meet the necessary transmission capacities, existing lines will be reinforced and new lines erected. The German energy agency calculated around 3,000 to 4,000 km of new transmission lines. Apart from permitting issues – where stakeholder groups, state authorities or even municipalities may actively block more centralized and faster decision mechanisms – and the "not-in-my-backyard" mentality of concerned residents, the question remains whether it makes economic sense to spend large amounts of money on a transmission system that may become obsolete in the near future.

The sequence of the steps being undertaken to enhance the energy transformation is not properly balanced, as Frank Hose from regional network operator ODR comments:

> We start by strengthening the grids, but in the long term less and less electricity will be transmitted. We concentrate on expanding decentralized generation plants but do not keep up with storage systems and smart grid systems. When intelligence is all finally in place, and we know at any time who is feeding in what and where, when we have weather forecasts and storage facilities, then we will not need powerful grids any longer. Electricity

will remain in the region. The expansion of substations and power lines may be important now, but not in the future. (Hose, ODR)

Positive network externalities related to the smart grid are diminished by too restrictive rate-of-return regulations of network operators and an incomplete assessment of the benefits of micro grids. In the long term, reinforcing the conventional grid is more expensive than a smart grid, according to EnBW Regional.

Government policy should be directed toward a least-cost approach to achieve CO_2 reduction targets. Building efficiency and energy performance contracting are the most effective means to obtain a carbon-free energy system. In addition, they enhance local value creation. But regulation currently favors subsidies for expensive technologies like photovoltaic plants and risks an expensive lock-in with a supply infrastructure more adapted to sunnier regions of the world. The German government has to come to terms with the fact that industrial policy to establish a high-wage country like Germany as a producer of photovoltaic panels has not yielded the expected benefits in terms of sustainable industry development.

9 From unbundling to rebundling

Rebundling may overcome inefficiencies in incentive schemes and information flows of the future smart grid

The political and academic debate about an optimal market design and sound regulatory framework in the electricity sector will surface again with the rise of a decentralized energy supply. The prevailing mantra of the separation of regulated network services in the transmission and distribution grid from competitive markets in generation and retail leads to information deficits and distorted incentives for grid control and optimization of power flows.

It is counterproductive if a sales division of a utility operates independently of the grid:

> The owner of a photovoltaic installation with a storage unit must be incentivized to feed his midday peak into the storage unit so that there can be savings in grid expansion. But if the sales department does not care about the grid and just wants to sell electricity to the customer, it has to adapt its strategy to the customer's wishes and allow him to consume electricity whenever he needs it. If the sales department is not able or allowed to develop a joint product with the grid operator, then we are lost. Grid and sales must sit together at the same table, in line with compliance. Gas supply, electricity supply, the communications systems, IT, grid – all parties must be involved. (Hose, ODR)

The coordination of unbundled market agents in the future energy grid will incur substantial transaction costs and the creation of multiple standards and protocols. Meanwhile, the information required to control the grid becomes increasingly valuable – and dispersed. Under these circumstances, it may not make sense to maintain the status quo of unbundling.

10 Public service obligation for transparency

Governments should be committed to tackle market barriers in building efficiency by promoting information platforms

The dispersed ownership structure of the buildings stock leads to inertia and the failure of incentive schemes beyond mere command-and-control mechanisms.

One way of overcoming market barriers in improving building efficiency is to provide adequate, specifically tailored information that is accessible to landlords and tenants. The new generation of digital natives will strongly rely on available web-based information.

If markets persistently fail, state authorities have the obligation to continue financing ventures like co2online, which provides consulting services for the initial assessment of residential energy consumption. To gain trust of the customers, these services have to be unbiased and should not be forced to generate revenue through alliances with crafts businesses and manufacturers of building efficiency devices.

Regulation has an important role to play in enhancing the transparency of options for renovation and conservation measures, and in funding organizations that provide these services.

Appendix: Company and Interviewee Profiles

Argentus

Argentus Energie is an independent energy service and consulting company in Germany. In the constantly changing and increasingly complex energy markets the company helps its clients to optimize their energy cost and to manage energy related risk. Argentus' services include high quality energy data management, market, commercial, and regulatory advisory as well as strategic energy sourcing services.

Many of the clients are real estate companies and/or have multiple sites where flexible energy contracts, reliable billing, and administration tools can add a lot of value. The Argentus team has reduced the energy cost of many thousands of buildings throughout Germany at two-digit rates.

Konrad Jerusalem

Dr Konrad Jerusalem is Managing Director of the energy consulting firm Argentus Energy. He received his law degrees at Ludwig Maximilians University in Munich and his doctorate at the Law Faculty of the University of Potsdam.

Prior to founding Argentus, Dr Jerusalem worked for seven years in international project finance at Dresdner Kleinwort in Frankfurt, Sao Paulo, and Rio de Janeiro. Thereafter, he joined RWE AG, where he served as deputy head of the operational M&A business. In 2008 he founded, together with Dr Georg Kofler, the energy performance contracting company Kofler Energies, where he served for about three years as an Executive Board Member.

Bioenergy village Jühnde

Jühnde was the first bioenergy village in Germany. The initial idea was to create a village that can supply itself with heat and power derived from bioenergy. The Multidisciplinary Centre for Sustainable Development of the University of Göttingen started the project in 2001. It took four years of preparation, construction work, and briefings until the bioenergy village Jühnde was established, including a CHP-plant with 700 kW electric power, a wood chips heating system with 550 kW thermal power, and a heat distribution network. At present, almost three quarters of the village inhabitants are members of the cooperative. The villagers have also established an information center for visitors.

Jühnde is located in the southern part of Lower Saxony and has around 800 inhabitants.

Eckhard Fangmeier

Eckhard Fangmeier studied physics at Georg-August University Göttingen. He works at a company for precision and control instruments and is responsible for quality and project management, energy management, and public relations. Since 2001 he is involved in the bioenergy village project Jühnde and founded the village initiative. In 2004, he became Chairman of the Board of the Cooperative and is Spokesman of the plant-operating company in Jühnde. Since 2010, he also serves as Managing Director of the Centre for New Energies and develops projects in the fields of renewable energies and electric mobility.

co2online

co2online is a non-profit limited liability company campaigning to reduce emissions of carbon dioxide (CO2) by means of energy saving, mostly through digital dialogue. co2online implements several campaigns for private households and communities, many co-funded by the German Federal Ministry for the Environment. The company offers online advisers on various aspects of room heating, energy saving modernization measures, and subsidies, as well as on various aspects of electricity saving. The online advisers are consulted by 10,000–15,000 people each week. Among other activities, co2online has compiled municipal heating surveys for 45 larger German cities, which are used to raise awareness of reasonable heating costs and heat consumption in centrally heated residential buildings.

The company has its headquarters in Berlin and employs around 40 people. It closely cooperates with its sister firm Senercon, which is in charge of the implementation of web services, software development and data analysis.

Johannes Dietrich Hengstenberg

Together with Hans-Peter Dürr, Dr Johannes Dietrich Hengstenberg founded in 1987 the Global Challenges Network. In 1990, he became co-founder and CEO of the Foundation for Environment and Nature Protection of the German Democratic Republic. In 1992, Hengstenberg initiated the Working Group Energy (Arbeitsgruppe Energie) in Munich, which developed into co2online (not-for-profit) and SEnerCon. For the campaign "Climate seeks protection," he won, with co2online, the Sustainable Energy Europe Award 2007 in the category Awareness Raiser in 2007. In 2008, he and co2online received the CleanTech Media Award. In 2009 his achievements were honored with the Merit Cross of the Federal Republic of Germany.

Daimler

Daimler AG is one of the biggest producers of premium cars, trucks, vans, and buses with headquarters in Stuttgart, Germany. The company achieved revenues of 107 billion Euro in 2011 and had a worldwide workforce of more than 270,000 employees.

The company's founders, Gottlieb Daimler and Carl Benz, made history with the invention of the automobile in the year 1886. Today, the company's portfolio of solutions ranges from the optimization of internal-combustion engines to hybrid drive and to locally emission-free driving. Daimler also responds to new mobility needs in the private and the public sectors with concepts such as car2go, a successful new car-sharing platform introduced in cities like Austin, Texas, and most recently Berlin. In 2013, the company plans to replace almost a third of its conventional smart fortwo car2go fleet in Berlin with smart fortwo electric drive models.

Ulrich Müller

Dr Ulrich Müller has been working with Daimler AG since 1985. After a trainee program he took over first responsibilities in Corporate Strategy and as assistant to the VP for Planning and Organization of the Passenger Car Development department.

After that he was in charge of various Daimler headquarter and divisional departments like Controlling, Organization, Investment Planning, Financial Services Strategy, and Vehicle Homologation as Senior Manager and Director.

He is now leading the Corporate Strategy department Global Regulatory and Innovation Strategy. Among others, this assignment encompasses worldwide responsibility for research strategy and a holistic Daimler strategy on sustainable products including the necessary overall processes. This assignment also includes Regulatory Strategy on all automotive issues worldwide.

E.ON

E.ON is one of the world's largest investor-owned power and gas companies. At facilities across Europe, Russia, and North America, nearly 79,000 employees generated just under €113 billion in sales in 2011.

E.ON was formed in June 2000 by the merger of VEBA and VIAG, two of Germany's largest industrial groups founded in the 1920s to serve as holding companies for state-owned industrial enterprises. E.ON AG in Düsseldorf oversees and coordinates the operations of five global units: Exploration & Production, Generation, New Build & Technology, Renewables and Optimization & Trading. Twelve regional units in Europe manage sales operations, regional energy networks, and distributed-generation businesses in their respective countries.

Eckhardt Rümmler

Eckhardt Rümmler completed his studies in Hamburg with an engineering degree in marine and shipbuilding technology in 1984. After that he built a broad operational and management expertise in several roles in the marine and energy industry. From 2005 to 2007 he was Member of the Management with E.ON Energie, München, with the responsibility for Energy and Business Optimization. He joined the E.ON Group Management in 2007 building up the new division for strategic steering of power generation and gas upstream.

Since 2010 he has been responsible as Senior Vice President for the E.ON's Group Strategy & Corporate Development, where he was one of the main contributors to the development of E.ON's new "cleaner & better" strategy in 2010.

EnBW Regional

EnBW Regional AG is the largest distribution network operator in the southwestern state of Baden-Wuerttemberg. The company is a subsidiary of German utility EnBW Energie Baden-Wuerttemberg AG and operates the high-, medium-, and low-voltage grids (110 kV, 20 kV, 0.4 kV) for its parent company. It has 2.95 million customers in electricity distribution and provides and sells network-related and community services for municipalities and public utilities in the electricity, gas, water, heat, and telecommunications. In addition, it operates the water network of the state capital Stuttgart, where it supplies approximately 600,000 citizens with drinking water.

EnBW Regional has about 3300 employees.

Michael Kirsch

Michael Kirsch studied industrial engineering with the specialization electrical engineering at the Technical University of Darmstadt. After having gained work experience in mergers and acquisitions at AMEOS group, he joined EnBW AG in 2004. In 2006, he entered the strategic asset management unit of the company's distribution network operations. In 2009, he joined the staff of the technical Executive Board and became director of a newly established division in charge of network development concepts in 2011. Since then, he is responsible for workforce management and the development of the smart grid at EnBW Regional AG.

GASAG

The business activities of the GASAG Group involve the transportation, distribution, and sale of natural gas, heat, power and water, the operation of storage facilities, the production of biomethane and the operation of facilities for distributed energy supply, consumption-billing and meter-reading services, meter management and the set-up, maintenance, repair, and overhaul of energy installations. The Group's core business is the transportation, distribution, and sale of energy and heat.

The GASAG Group's customers are mainly private household customers, trade and industry, hospitals, municipal entities, other gas supply companies, and gas transport services customers mainly in Berlin and surrounding Brandenburg.

The company has around 1900 employees. Revenues totaled €1.165 billion in 2011. As of December 2011, around 37 percent of GASAG's capital stock was held by E.ON Ruhrgas AG, Essen, and almost 32 percent by Vattenfall Europe AG, Berlin, and GDF SUEZ Beteiligungs GmbH, Berlin, respectively.

Andreas Prohl

Andreas Prohl holds a degree in industrial engineering with a specialization in mechanical engineering from the Technical University of Darmstadt. He joined BEB Erdgas und Erdöl GmbH in 1984 and worked there from 1988 in a leading management position. In 1991, he became Group Director Marketing of Wintershall Gas GmbH, and from 1993 onwards he was the company's Head of Department in the natural gas sales unit. After three years as Head of Business Development at WINGAS GmbH, he joined gas utility GASAG Berlin Gas AG and became Head of the Gas Procurement and Planning Department and later Head of the Sales and Marketing Department. Since 2002, he serves as Member of the Board of Directors with a focus on sales and engineering.

Greenvironment

Greenvironment is a European technology supplier and system integrator of power stations for decentralized power generation. The listed company with headquarters in Berlin was founded in 2002 in Finland and is now represented in Germany, the Czech Republic, Poland, and Romania. Greenvironment develops, builds, and operates natural gas or biogas-fired combined heat and power plants within the output segment of 50 kilowatt to 4 megawatt. The company currently operates more than 70 micro turbines in more than 30 plants with an average electrical output of 250 kilowatt.

The company is also an authorized distribution and service partner of the world leader for micro turbines, California-based manufacturer Capstone.

Apart from the planning, design and construction of turnkey CHP plants and the supply of components, Greenvironment also offers operator concepts for industry and local authorities. In the local authority sector it focuses on cooperations with municipal works.

Greenvironment has 33 employees and achieved net sales of more than €2 million in 2010.

Radu Anghel

In 2005 Radu Anghel started the development of the German business branch of Greenvironment GmbH, in charge of the development of the technical department and the after sales market of the company. In 2010, he was appointed managing director.

He is a 1994 graduate of the Technical University Timisoara (Romania) having a master's degree in mechanical engineering. He started his working career in the R&D departments of different German "hidden champions." After having received his MBA at Kellogg Management School of Chicago and WHU Vallendar, he led the internationalization of a producer of optical analytical instruments as international sales director before joining Greenvironment.

Itron

Itron is the world's leading provider of smart metering, data collection and utility software systems, with nearly 8,000 utilities worldwide relying on the company's technology to optimize the delivery and use of energy and water. The company is active in end-to-end smart grid and smart distribution solutions to electric, gas, and water utilities operating in over 130 countries. It has 105 million of its communication modules deployed and generated US-$2.4 billion in total revenue in 2011. On a global level, the company has over 9,000 employees.

Its German branch originated from AEG. It was taken over by Schlumberger. Through a leveraged buyout of Schlumberger Resource Management Services, which was part of Schlumberger Ltd, a global oilfield and information services company, Actaris was formed in 2001. In 2007, Itron acquired Actaris Metering Systems of Europe for more than US-$1 billion.

Karsten Peterson

Karsten Peterson holds a degree (FH) in the field of electrical engineering and automation technology. From 1976 to 1990 he worked at ZRW Oranienburg (now Deutsche Zählergesellschaft) in the meter control unit and later in the electronics and information technology department. In 1991 he joined Itron and worked from 1991 to 1996 at the company's Heliowatt plants in Berlin. Since 1996 he has been focusing on sales at Itron's Meter and Systems Engineering unit in Hameln. Since 2006, he is the company's sales manager for the division Electricity Germany.

Werner Paech

Werner Paech received vocational training as a merchant. In 1974, he started working for AEG in various functions of material management, including Head of Purchasing. In 1994 he joined the Schlumberger Group and became coordinator of the European purchasing activities of the company in 1998. He has held leadership roles of various purchasing and engineering competence teams. In 2002, he became Director of the operations of the electricity division of Germany. Since 2008 he has been serving as Managing Director for Electricity Germany; his responsibilities also include the Baltic States.

LichtBlick

With nearly 600,000 customers, LichtBlick is Germany's biggest independent provider of green energy. Founded just after the liberalization of the German electricity market in 1998, LichtBlick entered the mass market by offering "clean energy for a clean price," i.e. without nuclear and coal. Three quarters of the electricity are generated by hydroelectric power; the remaining 25 percent of electricity are produced with other renewable energy sources like solar and wind. In 2003, the company already counted 100,000 customers. In 2007, natural gas supply with five percent renewable biogas was added to LichtBlick's portfolio. Since then, more than 70,000 customers have chosen that option, making LichtBlick customers the largest consumer group for biogas in Germany. In 2010, the company launched its micro combined heat and power plant for residential customers together with car manufacturer Volkswagen.

LichtBlick AG is 100 percent privately owned, primarily by Hamburg merchants, and employs around 440 people.

Ralph Kampwirth

Ralph Kampwirth obtained a master's degree in politics at Bremen University. He started his professional career in 1997 at the citizens movement "More democracy." For several years he was a board member and from 1997 to 2004 Head of Communications. He has been Head of Corporate Communications and Politics of LichtBlick AG since April 2009. Between 2004 and 2009 he was Press Officer at World Wide Fund for Nature – WWF Germany – and was responsible for the communication of national and international environmental projects and campaigns.

ODR

EnBW Ostwürttemberg DonauRies AG is a regional utility active in the distribution of energy, natural gas, and water. It also operates in waste management, telecommunications, and other municipal services. In 2011, ODR supplied electricity to 124 municipalities with more than half a million inhabitants, and natural gas to 64 municipalities. The utility's service area is located in eastern Württemberg and in western Bavaria, southern Germany, and stretches over 3,400 square kilometers. Within this area, approximately 21,000 decentralized power generators, dominated by photovoltaic, wind, hydropower and biogas plants, feed their electricity into the grid of the utility.

Almost 90 percent of the shares of the company belongs to energy utility EnBW REG Beteiligungs GmbH.

In 2010, ODR had 334 employees and generated revenues of €424 million.

Frank Hose

Frank Hose has been working for 27 years in various positions of power and gas distribution for German energy utility Energie Baden-Württemberg AG (EnBW). After having graduated as an electrical engineer at the Technical University of Darmstadt, he began his professional career as a project engineer in the switchgear division of Brown, Boveri & Cie. AG in Mannheim. Since September 2008 he has been Executive Chairman of Ostwürttemberg Donau Ries AG in Ellwangen (ODR), a subsidiary of EnBW AG.

Siemens

With headquarters in Berlin and Munich, Siemens is a multinational player in the electrical engineering and electronics industry. Founded in 1847 by inventor and entrepreneur Werner von Siemens, the company had 360,000 employees and sales of €74 billion in 2011. Its business activities are focused on four sectors: Energy, Healthcare, Industry and Infrastructure & Cities.

The energy unit of the company has about 82,000 employees worldwide and provides solutions and technologies to generate, transmit, and distribute electrical power, both from fossil resources as well as solar, hydropower, and wind. Spanning the entire energy value chain, the company's product range also encompasses technologies for the production, conversion, and transport of primary fuels oil and gas. The Environmental Portfolio includes high-efficiency combined-cycle power plants, solutions for smart grids, electric mobility (from drives to rapid battery recharge stations), CO_2 separation for power plants, and power storage units for renewable energies.

Michael Weinhold

Michael Weinhold studied electrical engineering at Ruhr-University Bochum (Germany) and Purdue University, West Lafayette (USA). From 1989 to 1993, he worked as a research assistant for Power Electronics at Ruhr-University. After having joined Siemens AG in 1993, he started as a System Planning Engineer and was elected "Siemens Inventor of the Year" in 1997. He became the Commissioning Engineer for High Voltage Systems in 2002 and Siemens' CTO for Power Transmission and Distribution in 2004. Since 2008, he has been the CTO for Siemens in the field of Energy. In 2011, he became Member of the Siemens Sustainability Board. Weinhold also works as Adjunct Professor at Danmarks Tekniske Universitet (DTU), north of Kopenhagen.

Stadtwerke Krefeld

The SWK Stadtwerke Krefeld AG is a municipal utility and has been supplying electricity, gas, district heat and water to the German city of Krefeld and its surrounding region for 150 years. It also takes care of municipal sewage treatment and removal as well as waste management and local public transport.

SWK is involved in one of Germany's future energy grid projects, E-DeMa, which has been set up to develop and demonstrate locally networked energy systems. It allows via "smart gateways" to link private households to electricity suppliers and to develop an "energy market place of the future" where private households can interact with energy dealers or energy distribution system providers to market their decentrally produced power. The project also includes trials with electrical vehicles used for flexible energy storage. It is a joint project of the municipal utility with energy incumbent RWE, electronics manufacturer Siemens and other partners.

SWK supplies around 155,000 inhabitants with electricity.

Carsten Liedtke

Carsten Liedtke began his career at the division of private communication systems and networks of Siemens AG in Dortmund and Munich. He then took a position as assistant of the Management Board of VIAG GmbH & Co. KG (now O_2) in Munich. From 1999 to 2003, he worked at the RWE Group as Manager for investor relations in charge of communication with institutional investors. At RWE Rhein-Ruhr AG, Mr Liedtke became Head of Corporate Development and served temporarily as Acting Managing Director of RWE Customer Service GmbH. Since September 2007, he has been Member of the Board of Directors of the municipal utility SWK Stadtwerke Krefeld AG. In November 2011 he became Speaker of the Board of Directors and is responsible for energy and transport in the utility.

Stadtwerke Unna

The municipal utility SWU has 35,000 energy customers and offers electricity, gas, and heat. Although it purchases the bulk of its energy from outside providers, it started in the 1990s to implement environmentally friendly self-production facilities like five cogeneration plants, a wind turbine and photovoltaic systems. In addition, the company engages in consulting services regarding energy efficiency for households and businesses, and energy performance contracting. In particular, the utility has introduced an environmental assessment of local businesses. Together with private consultants and specialist advisers, the utility reviews opportunities for environmental and economic improvements in the respective firms and examines the possibilities for integrated energy management systems.

Around 76 percent SWU is owned by the city of Unna GmbH, and 24 percent belongs to energy utility RWE.

Christian Jänig

Prof Dr Christian Jänig studied business administration at the University of Mannheim and received his doctorate at the German University of Administrative Sciences, Speyer. After academic positions at the universities of Mannheim and Munich, he started working for MVV mbH, the municipal utility in Mannheim, in 1976. Between 1980 to 1990 he held CEO positions at the municipal utility Achim and at the gas supplier Langwedel. In 1990, he became CEO of the municipal utility Unna. In 2002, he was appointed to the Board of Directors of the German Association of Local Utilities (VKU). Jänig became Honorary Professor at the Bremerhaven University of Applied Sciences in 1995 and has been teaching topics like system analysis in public administration, economics, business decision processes, and risk management.

Viessmann

The Viessmann Group is one of the leading international manufacturers of heating systems. Founded in 1917, the family business maintains a staff of approximately 9,600 employees and generates €1.86 billion in annual group turnover. Viessmann is an internationally orientated company with 24 production divisions in 11 countries. Fifty-five percent of sales are derived from export activities.

Viessmann offers system solutions with heat sources for all fuel types and outputs ranging from 1.5 to 116,000 kW. Viessmann's product portfolio includes combined heat and power stations for oil or bio natural gas, with outputs from 1 to 401 kW_{el} and 6 to 549 kW_{th}, systems for renewables including solar thermal and photovoltaic systems, boilers and combustion units for biomass and heat pumps. In its Efficiency Plus project it showed that the German federal government's energy and climate policy targets for 2050, namely a reduction in primary energy demand by 50 percent relative to 2008, an increase in the share of renewable energies to 60 percent as well as a reduction in CO_2 emissions by 80 percent relative to 1990, are attainable today with the help of technology that is already available in the market. For this project, Viessmann received the German Sustainability Award 2009 in the category Most Sustainable Production as well as the German Energy Agency's Efficiency Award in 2010.

Manfred Greis

Manfred Greis studied to become a secondary school teacher, but started a career at the Viessmann Werke in Allendorf (Eder) in 1981. After holding various positions in sales, he is now in charge of corporate communications, political contacts, the involvement of the company in associations, and the sustainability strategy of the Viessmann Group. He also holds a number of honorary positions in national and international organizations, including Vice President of the Federal Industrial Association of Germany – House,

Energy and Environmental Technology (BDH) and Member of the Executive Council of the Association of the European Heating Industry (EHI) and the Environmental and Energy Committee of German Chamber for Industry and Commerce (DIHK).

VW

The Volkswagen Group, with its headquarters in Wolfsburg, is one of the world's leading automobile manufacturers and the largest carmaker in Europe. The Group operates 94 production plants in 18 European countries and an additional eight countries in the Americas, Asia, and Africa. In 2011, the Group delivered 8.265 million vehicles, corresponding to a 12.3 percent share of the world passenger car market. With over 500,000 employees, sales revenue in 2011 totaled €159 billion.

The Group is made up of ten brands from seven European countries, including Volkswagen, Audi, SEAT, ŠKODA, Bentley, Bugatti, Lamborghini, Volkswagen Commercial Vehicles, Scania, and MAN. The Volkswagen Group is also active in other fields of business, manufacturing large-bore diesel engines for marine and stationary applications (turnkey power plants), turbochargers, turbo machinery (steam and gas turbines), compressors and chemical reactors, and also producing vehicle transmissions, special gear units for wind turbines, slide bearings, and couplings as well as testing systems for the mobility sector.

Jürgen Willand

Dipl.-Ing. Jürgen Willand has been working in the Research & Development Department in Wolfsburg before taking over the responsibility for the development of the micro CHP production line of Volkswagen in Salzgitter.

Younicos

Younicos is a startup founded in 2008 and focuses on the application of energy storage and grid management technologies. The company plans, constructs, and operates autonomous energy systems using up to 100 percent clean energy to supply islands or remote areas with power. It sells combined power-storage systems for large continental grids that store excess electricity and distributed storage systems that allow regionally generated renewable energy to be consumed where it is generated. In addition, Younicos provides analysis and consultancy services on efficient energy consumption and building design, and develops customized energy supply solutions on the basis of renewable energy sources.

The company's energy system projects include one of the world's first fully self-sustaining zero-emissions energy-efficient buildings and a large-scale, completely self-sufficient power supply and grid network based on up to 100 percent wind, solar, and biomass and megawatt-scale batteries.

The company is headquartered in Berlin and currently has a staff of 50 people.

Alexander Voigt

Alexander Voigt obtained his degree in physics and since 1986 has been an active player in the photovoltaics industry with his own companies. In 1997, he co-founded SOLON AG, previously one of the leading European companies manufacturing solar modules and photovoltaic power plants. Two years later, Alexander Voigt was one of the driving forces behind the spin-off of SOLON's production technology business unit into the newly founded Q-Cells AG. From 2001 to 2006, he was Chairman of the Executive Board of SOLON AG, moving to the Supervisory Board in 2006. In 2008 he and his business partner Clemens Triebel together founded Younicos AG in Berlin as a merger of the two companies SOLON Laboratories AG and I-Sol Ventures GmbH. As CEO of Younicos AG, Alexander Voigt also takes on special responsibility for the Sales and Marketing divisions.

Notes

Introduction

1. The quotes represent the interviewees' personal opinions and may not necessarily reflect their respective companies' viewpoints.
2. Developing countries with a patchy grid, insufficient electrification, and frequent system outages are, of course, the most obvious candidates for a rapid deployment of decentralized energy, because system inertia and an urge for change is less pronounced than in industrialized countries.

1 Empowerment Paradigm – The Age of the Prosumer

1. See, for example, https://www.genossenschaftsverband.de/verband/presseservice/pressemeldungen-aus-dem-verband/energieversorgung-autark-und-selbstbestimmt.

2 Small Is Beautiful

1. The business model of LichtBlick and VW will be presented in detail in the section "Exploiting scarcity pricing in the wholesale market".

3 The Rise of Island Systems

1. See World Bank (2002), *Empowerment and Poverty Reduction: A Sourcebook*.
2. See Chapter 6 for a more detailed discussion of this topic.

5 Local Storage Solutions

1. Stuxnet was a computer virus that appeared in 2010 and was able to attack the control systems of complex machinery and industrial processes, including the treatment of nuclear materials.

6 Enabling Negawatts

1. The ICF comments on the third constraint: "The landlord agencies, the GSA [the government's landlord, providing office and other workspace services for the federal government] at the federal level and its equivalent in the various states, have spent more than a century getting control of the buildings in their domains. EPC [Energy Performance Contracting] is a disruptive project model to these agencies, because it displaces longstanding contract methodologies and contractor relationships. The financial control agencies, such as the OMB or individual agency comptrollers at the federal level, are generally unfamiliar with, and suspicious of, the economics of EPC, and so resist its widespread implementation" (ICF 2007, 33).

2. "A white certificate is an instrument issued by an authority or an authorised body providing a guarantee that a certain amount of energy savings has been achieved. Each certificate is a unique and traceable commodity that carries a property right over a certain amount of additional savings and guarantees that the benefit of these savings has not been accounted for elsewhere," as defined in Bertoldi and Rezessy (2009, 25).

7 Insights from Germany for a Decentralized Energy Future

1. Aristotle, *Metaphysica*.

References

Andre, L. (2010). Søren Hermanson – Bio, *Sustainability Pay$ – The Copenhagen Experience*. Rice University, Houston, Texas: Shell Center for Sustainability.

Anthony, S. (2012). IBM Creates Breathing, High-Density, Light-Weight Lithium-Air Battery. Retrieved June 30, 2012, from http://www.extremetech.com/

Arthur D. Little (1981). *The Strategic Management of Technology*. Cambridge, MA: Arthur D. Little.

Arup (2012). *Druk White Lotus School*. Retrieved June 30, 2012, from http://www.arup.com/Projects

Auer, J., & Keil, J. (2012). *State-of-the-Art Electricity Storage Systems*. Frankfurt am Main, Germany: Deutsche Bank.

Auraplan (2009). Types of Energy Efficient Buildings. In J. Faltin & C. v. Knorre (Eds.), *Energieeffizienz von Gebäuden*, Hamburg: Auraplan.

Auto-Medienportal.Net/Volkswagen (2012). *Die Serienproduktion von Zuhause-Kraftwerken im Volkswagen-Motorenwerk Salzgitter*. Retrieved June 30, 2012, from http://www.auto-medienportal.net/

AWEA (2010). *AWEA Small Wind Turbine Global Market Study*. Washington, DC: American Wind Energy Association.

AWEA (2012). *2011 U.S. Small Wind Turbine Market Report*. Washington, DC: American Wind Energy Association.

Bauknecht, D. (2012). *Transforming the Grid: Electricity System Governance and Network Integration of Distributed Generation*. Baden Baden: Nomos.

Bebchuk, L. A. (1992). Federalism and the Corporation: the Desirable Limits on State Competition in Corporate Law. *Harvard Law Review*, 105(7), 1435–1510.

Bertoldi, P., & Rezessy, S. (2009). *Energy Saving Obligations and Tradable White Certificates*. Brussels: Joint Research Centre of the European Commission.

Bewarder, M. (2011). Wenn Bürger über den Haushalt mitentscheiden. *Welt Online*, July 12.

Bioenergy village Jühnde (2012). *Das Bioenergiedorf Jühnde - Modelldorf für eine eigenständige Wärme- und Stromversorgung*. Permission granted by Eckhard Fangmeier on July 2, 2012.

Booth, A., Demirdoven, N., & Tal, H. (2010). The Smart Grid Opportunity for Solutions Providers. In D. Mark, K. Ostrowski & H. Tal (Eds.), *Can the Smart Grid Live Up to Its Expectations?* (Vol. 1). New York City: McKinsey.

Borbely, A. M., & Kreider, J. F. (2001). *Distributed Generation: The Power Paradigm for the New Millennium*. Boca Raton, FL: CRC Press.

BPIE (2011). *Europe's Buildings Under the Microscope: Country-by-Country Review of the Energy Performance of Europe's Buildings*. Brussels: Buildings Performance Institute Europe.

Brown, M. A. (2001). Market Failures and Barriers as a Basis for Clean Energy Policies. *Energy Policy*, 29(14), 1197–1207.

Bundesnetzagentur (2011). *Markt und Wettbewerb: Energie Kennzahlen 2010*. Bonn: Bundesnetzagentur.

BWP (2011). *BWP-Branchenstudie 2011*. Berlin: Bundesverband Wärmepumpe.

Carrick, A.-M. (2009). Alumnus Entrepreneur Profile: Sameer Hajee. Retrieved June 30, 2012, from http://www.insead.edu

Cary, W. L. (1974). Federalism and Corporate Law: Reflections upon Delaware. *Yale Law Journal*, 83(4), 663–705.

Chamra, L. M., & Mago, P. J. (2009). *Micro-Chp Power Generation for Residential and Small Commerical Buildings*. Commack, NY: Nova Science.

Christensen, C. M. (1997). *The Innovator's Dilemma: When New Technologies Cause Great Firms to Fail*. Cambridge, MA: Harvard Business School Press.

Christensen, C. M., & Rosenbloom, R. S. (2000). Explaining the Attacker's Advantage: Technological Paradigms, Organizational Dynamics, and the Value Network. In D. B. Audretsch & S. Klepper (Eds.), *Innovation, Evolution of Industry and Economic Growth. Volume 3* (pp. 125–149). Cheltenham and Northampton, MA: Edward Elgar.

COGEN Europe (2011). *Cogeneration 2050 – The Role of Cogeneration in a European Decarbonised Energy System*. Brussels: COGEN Europe.

Cotti, M. (2008). Smart Meters – The Technical Potential and the Commercial Opportunities, Florence: European University Institute.

co2online (2012). *Share of German heating systems replaced annually*. Personal communication on June 29, 2012.

De Wachter, B. (2006). Japanese Top Runner Program. Retrieved June 30, 2012, from http://www.leonardo-energy.org

Deans, G. K., Kroeger, F., & Zeisel, S. (2002). *Winning the Merger Endgame: A Playbook for Profiting From Industry Consolidation*. Maidenhead: McGraw-Hill.

DeCanio, S. J. (1993). Barriers within Firm to Energy-Efficient Investments. *Energy Policy*, 21(9), 906–914.

Delta-ee (2012). *Micro-CHP Sales by Technology Type*. Personal communication by S. Dwyer, received July 5, 2012.

dena (2007). *Contracting-Potenzial in öffentlichen Liegenschaften*. Berlin, Germany: Deutsche Energie-Agentur.

dena (2012). E.ON Ruhrgas Pilotanlage, Falkenhagen. Retrieved June 30, 2012, from http://www.powertogas.info/

Ding, J., & Huihuang, C. (2012). *Case Study for the Distributed Power Supply System*. Shanghai: CELAP.

Droege, P. (2009). *100% Renewable: Energy Autonomy in Action*. London: Earthscan.

Druk White Lotus School (2012). *Sustainable Design Examples*. Retrieved June 30, 2012, from http://www.dwls.org

DZWWW (2003). Dezhou. Retrieved June 30, 2012, from http://www.dzwww.com

e-Business Watch (2009). *Metering and Measurement Facilities as Enabling Technologies for Smart Electricity Grids in Europe*. Brussels: European Commission, Enterprise & Industry Directorate General.

Fehrenbacher, K. (2010). Smartgridcity Is a Smart Grid Flop. *GigaOM*, August 4.

Feldman, A. (2011). For Electricity 2.0, a Short Circuit in Boulder. *Time*, January 25.

Frost & Sullivan (2011a). *Frost & Sullivan Confirms 81% of Smart Meters Installed in Europe are Based on Echelon Technology* Mountain View, CA: Frost & Sullivan.

Frost & Sullivan (2011b). Smart Meter Market in Europe Heating Up. *Smartgridnews.com*, July 20.

Galvin, R., & Yeager, K. (2008). *Perfect Power: How the MicroGrid Revolution Will Unleash Cleaner, Greener, More Abundant Energy*. Maidenhead: McGraw-Hill.

General Electric (2012). *Smarter GE Appliances for a Smarter Home*. Retrieved on June 30, 2012, from http://www.gereports.com

Ghemawat, P. (1991). *Commitment: The Dynamic of Strategy*. New York City, NY: Free Press.

Gore, A. (2009). *Our Choice: A Plan to Solve the Climate Crisis*. London: Bloomsbury.
Greenvironment (2012). *Biogas plant Muntscha*. Retrieved June 30, 2012 from http://greenvironment.de
Grossman, S. J., & Hart, O. D. (1986). The Costs and Benefits of Ownership: A Theory of Vertical and Lateral Integration. *Journal of Political Economy, 94*(4), 691–719.
Hannan, M. T., & Freeman, J. (1984). Structural Inertia and Organizational Change. *American Sociological Review, 49*(2), 149–164.
Hart, O. D. (1995). *Firms, Contracts, and Financial Structure*. Oxford: Clarendon Press.
Hart, O. D. & Moore, J. (1990). Property Rights and the Nature of the Firm. *Journal of Political Economy, 98*(6), 1119–1158.
Hartmann, G. (2012). Wenn sich Häuser den Wetterbericht zunutze machen. *Welt Online*, May 19.
Henderson, R. (1993). Underinvestment and Incompetence as Responses to Radical Innovation: Evidence from the Photolithographic Alignment Equipment Industry. *RAND Journal of Economics, 24*(2), 248–270.
Hill, C. W. L., & Rothaermel, F. T. (2003). The Performance of Incumbent Firms in the Face of Radical Technological Innovation. *Academy of Management Review, 28*(2), 257–274.
Hofmann, M. (2012). Abwrackprämie soll Einbau neuer Heizkessel fördern. *Welt Online*, February 6.
Honda (2009). *Vaillant and Honda Develop Cogeneration System for Single Family Homes*. Retrieved June 30, 2012, from http://world.honda.com
ICER (2012). *Smart Meters in Italy*. London: International Confederation of Energy Regulators.
ICF & NAESCO (2007). *Introduction to Energy Performance Contracting*. Washington, DC: Environmental Protection Agency.
IEA (2007). *Mind the Gap – Quantifying Principal-Agent Problems in Energy Efficiency*. Paris: International Energy Agency.
IEA (2008). *Combined Heat and Power: Evaluating the Benefits of Greater Global Investment*. Paris: International Energy Agency.
IEA (2009). *Cogeneration and District Energy*. Paris: International Energy Agency.
IEA (2011). *Technology Roadmap: Smart Grids*. Paris: International Energy Agency.
Indian Ministry of New and Renewable Energy (2012). *Druk Padma Karpo School*. Retrieved June 30, 2012, from http://ncict.net/Examples
IWU (2007). Modellprojekt "Ökologischer Mietspiegel Darmstadt", Phase III Retrieved June 30, 2012, from http://www.iwu.de
Kaplan, S. M., Sissine, F., Abel, A., Wellinghoff, J., Kelly, S. G., & Hoecker, J. J. (2009). *Smart Grid: Modernizing Electric Power Transmission and Distribution*. Alexandria, VA: TheCapitol.Net.
Kolanowski, B. F. (2008). *Small-Scale Cogeneration Handbook*. Lilburn, GA: Fairmont Press.
Kuhn, T. S. (1962). *The Structure of Scientific Revolutions*. Chicago, IL: University of Chicago Press.
Laughlin, R. B. (2011). *Powering the Future: How We Will (Eventually) Solve the Energy Crisis and Fuel the Civilization of Tomorrow*. New York City: Basic Books.
Leibenstein, H. (1966). Allocative Efficiency vs. "X-Efficiency". *American Economic Review, 56*(3), 392–415.
Liebowitz, S. J., & Margolis, S. E. (1994). Network Externality: An Uncommon Tragedy. *Journal of Economic Perspectives, 8*(2), 133–150.
Lopatka, J. E., & Page, W. H. (1999). Network Externalities. In B. Bouckaert & G. De Geest (Eds.), *Encyclopedia of Law and Economics*. Cheltenham: Edward Elgar.

Lorenz, G., & Mandatova, P. (2011). EURELECTRIC's 10 Steps towards Smart Grids with a Look at Flexible Loads. *European Heat Pump NEWS*, –June 2.

Markard, J., & Truffer, B. (2006). Innovation Processes in Large Technical Systems: Market Liberalization as a Driver for Radical Change? *Research Policy*, 35(5), 609–625.

Masdar City (2011). What Is Masdar City? Sustainability and the City. Retrieved June 30, 2012, from http://www.masdarcity.ae

Mayer, A. (2010). *Energy Performance Contracting in the eu: Introduction, Barriers and Prospects*. Washington, DC: Johnson Controls.

Masdar (2012). *Proposed Master Plan of Masdar City*. Retrieved June, 30, 2012, from http://www.masdar.ae

Mayntz, R. (1993). Modernization and the Logic of Interorganizational Networks. In J. Child, M. Crozier & R. Mayntz (Eds.), *Societal Change Between Market and Organization*. Aldershot: Avebury.

McKinsey (2007). *Costs and Potentials of Greenhouse Gas Abatement in Germany*, Berlin: McKinsey.

Meier, A., & Whittier, J. (1983). Consumer Discount Rates Implied by Purchases of Energy-Efficient Refrigerators. *Energy Policy*, 8(12), 957–962.

MeteoViva (2012). Energieeffizienz mit Wettervorhersage-Steuerung. Retrieved June 30, from http://www.meteoviva.com/

Mukherjee, J. (2008). Societal Benefits of Smart Grid – An Economics Perspective. *KEMA Automation Insight Newsletter* (April).

Müller, C., Wissner, M., & Growitsch, C. (2010). *The Economics of Smart Grids*. Brussels: CRNI.

Narayan-Parker, D. (2002). *Empowerment and Poverty Reduction: A Sourcebook*. Washington DC: World Bank.

Norris, T. (2010). China Building Ambitious "Solar Valley City" to Advance Solar Industry *Huffington Post*, April 18.

Nuru Energy (2012). Why Have Other Solutions Failed? Retrieved June 30, 2012, from http://nuruenergy.com

O'Reilly III, C. A., & Tushman, M. L. (2004). The Ambidextrous Organization. *Harvard Business Review*, 82(4), 74–81.

Page, N., & Czuba, C. E. (1999). Empowerment: What Is It? *Journal of Extension*, 37(5).

Pahl, G. (2007). *The Citizen-Powered Energy Handbook: Community Solutions to a Global Crisis*. White River Junction, VT: Chelsea Green.

Pehnt, M., Cames, M., Fischer, C., Praetorius, B., Schneider, L., Schumacher, K., et al. (2010). *Micro Cogeneration: Towards Decentralized Energy Systems*. Wien/New York: Springer.

Pike Research (2010a). *Energy Efficiency Retrofits for Commercial Buildings Could Save $41.1 Billion per Year in Energy Costs*. Boulder, CO: Pike Research.

Pike Research (2010b). *Smart Grid Data Analytics Market to Reach $4.2 Billion by 2015*. Boulder, CO: Pike Research.

Pike Research (2011). *Ultracapacitor Market to Top $900 Million by 2016*. Boulder, CO: Pike Research.

Porter, M. E. (1980). *Competitive Strategy: Techniques for Analyzing Industries and Competitors*. New York City, NY: Free Press.

Quiet Revolution (2011). UK & Ireland market. Retrieved June 30, 2012, from http://www.quietrevolution.com

REGBIE+ (2009). *European Biomass Success Stories: Policy and Legislation – Bioenergy Village Jühnde*. Hannover: Regional Initiatives Increasing the Market for Biomass Heating in Europe.

Research In China (2011). *China Smart Meter Industry Report, 2011–2012*. Beijing: Research In China.

Rifkin, J. (2011). *The Third Industrial Revolution: How Lateral Power Is Transforming Energy, the Economy, and the World*. Basingstoke: Palgrave Macmillan.

Ross, S. (2004). Energieeinspar-Contracting – Ein Auslaufmodell? *E-world*. Essen, Germany.

Rothaermel, F. T. (2001). Complementary Assets, Strategic Alliances, and the Incumbent's Advantage: An Empirical Study of Industry and Firm Effects in the Biopharmaceutical Industry. *Research Policy, 30*(8), 1235–1251.

Samso Energy Agency (2012). Current Projects. Retrieved June 30, 2012, from http://www.seagency.dk

Schlandt, J. (2012a). Energie aus dem Keller. *Frankfurter Rundschau*, March 26.

Schlandt, J. (2012b). Eine heiße Verbindung. *Frankfurter Rundschau*, April 14.

Schumpeter, J. (1934). *The Theory of Economic Development*. Cambridge: Harvard University Press.

Sen, A. (2003). The Possibility of Social Choice. In T. Persson (Ed.), *Nobel Lectures, Economics 1996–2000* (pp. 178–215). Singapore: World Scientific Publishing.

smartmeters.com (2011). Landis+Gyr Maintains Its Leading Global Position. Retrieved June 30, 2012, from http://www.smartmeters.com

So, K., & Lagerling, C. (2009). *How Real Is the Vision of a "Smart Grid"?* San Francisco: GP Bullhound.

Solar Cities Initiative World Congress. (2010). Solar Industry of Dezhou. Retrieved June 30, 2012, from http://www.chinasolarcity.cn

Stiebel Eltron (2010). Die Zukunft der dezentralen Energiegewinnung. Retrieved June 30, 2012, from http://www.stiebel-eltron.de

Stieß, I., van der Land, V., Birzle-Harder, B., & Deffner, J. (2010). *Handlungsmotive – hemnisse und Zielgruppen für eine energetische Gebäudesanierung*. Frankfurt am Main: ENEF Haus.

Stiglitz, J. E. (2002). *Globalization and Its Discontents*. New York City: W.W. Norton.

Su Li/Greenpeace (2010). *Solar Thermal Water Heaters on the Roof of Residential Buildings in Dezhou*. Retrieved on June 30, 2012, from http://www.greenpeace.org

Sull, D. N., Tedlow, R. S., & Rosenbloom, R. S. (1997). Managerial Commitments and Technological Change in the US Tire Industry. *Industrial and Corporate Change, 6*(2), 461–501.

Sweatman, P., & Managan, K. (2010). Financing Energy Efficiency Building Retrofits. IE Business School: Instituto de Empresa.

Swierczynski, M. J., Teodorescu, R., Rasmussen, C. N., Rodriguez, P., & Vikelgaard, H. (2010). Overview of the Energy Storage Systems for Wind Power Integration Enhancement, *IEEE International Symposium on Industrial Electronics (ISIE)*. Bari: IEEE Press.

Tanaka, N. (2008). Today's Energy Challenges: The Role of CHP. In C. Europe (Ed.), *Cogen Europe Annual Meeting*. Brussels: Cogen.

Tirole, J. (1986). Procurement and Renegotiation. *Journal of Political Economy, 94*(2), 235–259.

Tokyo Gas (2011). Development of the New Model of a Residential Fuel Cell, "ENE/FARM". Retrieved 30 June, 2012, from http://www.tokyo-gas.co.jp

Tokyo Gas (2012). Residential Fuel Cell "Ene-Farm" Reaches 10,000 Units Sold on a Cumulative Basis. Retrieved June 30, 2012, from http://www.tokyo-gas.co.jp

Tokyo Gas, & Panasonic (2011). Tokyo Gas and Panasonic to Launch New Improved "Ene-Farm" Home Fuel Cell with World-Highest Power Generation Efficiency. Retrieved June 30, 2012, from http://panasonic.co.jp

UK Department of Energy and Climate Change (2012). Efficiency Gains of Cogeneration, Compared to Conventional Methods. Retrieved June 30, 2012, from http://chp.decc.gov.uk/cms/

United Nations (2012). *United Nations E-Government Survey 2012: E-Government for the People*. New York City: UN.

Uttich, S. (2010). Tauwetter für eingefrorene KfW-Förderprogramme. *FAZ*, January 20.

Vidal, J. (2011). Masdar City – A Glimpse of the Future in the Desert. *Guardian*, April 26.

Vogel, D. (1997). *Trading Up: Consumer and Environmental Regulation in a Global Economy*. Cambridge, MA: Harvard University Press.

von Weizsäcker, E.-U., Lovins, A. B., & Lovins, L. H. (1998). *Factor Four: Doubling Wealth – Halving Resource Use. The New Report to the Club of Rome*. London: Earthscan.

von Weizsäcker, E.-U., Hargroves, K., Smith, M. H., Desha, C., & Stasinopoulos, P. (2009). *Factor Five: Transforming the Global Economy Through 80% Improvements in Resource Productivity – A Report to the Club of Rome*. London: Earthscan.

WADE (2010). *The Potential for Clean DE and CHP in China*. Washington, DC: World Alliance for Decentralized Energy.

WADE (2012). Microturbines. Retrieved June 30, 2012, from http://www.localpower.org

WBCSD (2009). *Transforming the Market: Energy Efficiency in Buildings*. Geneva: World Business Council for Sustainable Development.

Weinmann, J. (2007). Agglomerative Magnets and Informal Regulatory Networks – Electricity Market Design Convergence in the USA and Continental Europe. Florence: European University Institute.

Williamson, O. E. (1985). *The Economic Institutions of Capitalism: Firms, Markets, Relational Contracting*. New York City: Free Press.

Xcel Energy (2012). What Is SmartGridCity? Retrieved June 30, 2012, from http://smartgridcity.xcelenergy.com/

Key concepts, persons and technologies

abatement costs, 147, 150, **151**, 152, 160, 161
ambidextrous organizations, **19**

B2B, **56**
B2C, **56**
Bakke, Dennis W., 10, 12
batteries, 13, 80, 92, **131–3**, 134, **136–45**, 146, 147, 149
bioenergy village, 3, **68–80**, 86, 186
 see also biogas; empowerment; Jühnde
biogas, 24, **52–8**, 59–60, 68, 69, 71–2, 101
Boulder, **105**
building efficiency, 3, 4, 5, **158–64**, 165, 169–71, 183, 192, 193

California effect, **67**
 see also race to the top
China, 5, 22, 66, 85, 86, 90, 116, **143**, 148, 158, 170
CHP, *see* combined heat and power (CHP)
Christensen, Clayton, **1**
citizen value, **80–4**
climate change, 6, 7, 13, 14, 20, 164
cogeneration, 2, 3, 5, 14, **19**, 20, 21–2, 24, 25, 30, 33, 38, 39, 42, 45, 51, 92, 171, 180
 see also combined heat and power (CHP)
combined heat and power (CHP), 5, **19**, **32–5**, **55–6**, 59, 69, 72, 188
competitive advantage, 2, **18**, 24, 56, 90, **95**, 121, 145, 189
contracting, *see* energy performance contracting
creative destruction, 5, **18**

Delaware effect, *see* race to the top
digital natives, 65, 155, 183, 193
disruptive innovations, 1, 18, **19**, 186

economies of scale, 8, 11, **12**, **17–18**, 41, 67, 88, 115, 141, 187
economics paradigm, 9, 10, 11, **12**, 15
Edison, Thomas Alva, **9**, 12
efficiency gap, **151**, 156, 162
electric vehicles, 13, 71, **76**, 84, 98, 105, 107, 115, 131, **140–9**
empowerment, 7, **12**, **14**, **64–6**
empowerment drivers,
 see empowerment
empowerment paradigm, *see* empowerment
energy performance contracting, 4–5, 109, 118, 152, **157**, **169–83**, 189–90, 192, 207, 211
energy policy triangle, **1–2**
energy trajectory, **8**
engineering paradigm, 9, 10, **12**, 15
European Union, 11, 12, 23, 65, 98, 113, 127, 135, 144, 153, **167**, 173

Friedman, Milton, **10**, 12
fuel cells, **21**, 23, 26–7, **132**, 141, 143–4, 177–8
Fukushima, 3, 24

global warming, 7
 see also climate change
globalization, 12, **14**, 66, 187

Hajee, Sameer, 12, **15**, 62, 63
Hermansen, Søren, 12, 15, **69–71**
hydroelectric power, 7, 9–10, 101, 203
 see also renewable energies

ICT, *see* information and communication technology (ICT)
incomplete contracts, **156–7**, 180
incumbent, 1, 2, 3, 10, 12, **18–19**, 31, 37, 38, 40, 69, 75, 90, 118, 148, 186, 187, 188, 206

219

India, 5, 63, 80, 158, **170–1**
industry consolidation, 11, 12, **18**, 90
industry phases, **18**
information and communication technology (ICT), 10–11, 12, **13**, 105, 109, **122**, 128, 129, 175, **181–2**, 189, 190
intelligent meter, *see* smart meter

Jühnde, 3, **68–80**, 191, 195–6

Kuhn, Thomas, 15

large technical systems, 2, **9**, 13, 185
lock-in, 39, **89–91**, 114, 181, 183, 192

market barriers, 5, 151, **153–5**, 161, 164, 183, 189, 193
market failures, 5, 13, 90, 151, **155–6**, 157
Masdar, **79–80**
municipal utilities, 3, 40, 58–9, 63, 67, **80–4**, 93, 102, 103, 107–8, 109–10, 119, 127, 129, 144, 157, 176–7, 178, 185, 187, 189, 206, 207

network externalities, **88–9**, 90, 98, 119, 120, 187, 192

open innovation, **96**, 191
organizational designs, **19**

paradigm, **9**
photovoltaics, 1, 4, 8, 11, 30, 35, 41, 69, 79, 80, 85, **93–4**, 96, 99–102, 105, 106–7, 110, 115, 121, 122, 127, 132, 133–4, 136, 137–8, 148, 151, 161, 170, 186, 192
Porter, Michael, 67
principal-agent dilemma, 151, **155–6**
prosumer, 7, **15**, 92, **121–2**, 129, 189, 191
Purpa Act, 11

race to the top, **66–7**, 148, 149
Reagan, Ronald, 10, 12, 15
recommunalization, 4, 12, **80–4**, 108
regulation, 3, 4, 10, 12, 14, 25, 66, 72, 73, 87, 104, 106, 107, 109, 111, 113, 114, 115, 127, 128, 135, 140, **146**, 153, 154, 156, 173, 183, 185, 191, 193

regulatory drivers, **13–14**
see also regulation
renewable energies, 2, 8–9, 11, 14, 15, 23, 24, 25, 30, 32–3, 34, 41, 52, 59, 69–70, 71, 72, 80, 83, 87, 88, 89, 92, 94–106, 115, 116, 120, 124, 127, 131, 133, 134, 135, 137–8, 140, 158, 159, 161–2, 176, 185, 188, 195, 203, 205, 208, 209
Rifkin, Jeremy, 6, **13**

Samsø, **70–1**
Sant, Roger W., 10, 12
Schumpeter, Joseph, **18**
security of supply, 1–2, 13, 14, 76, **94**, 97, 174
Sen, Amartya, **64–5**
smart grid, 4, 14, 78, 87, 88–9, 90, **91–3**, 94, 95, 97–8, 99, 101, 104, **105–10**, 118, 120, 124, 126, 127, 128, 136, 140, 158, 159, 189, 191, 192, 199, 202, 205
smart home, **125–7**
smart meter, 2, 3, 13, 87, 88, 89, 90–1, 105, 106, 107–8, 109, 110–18, **119–24**, 125, 126, 127–8, 129, 181, 187, 189, 202
split incentives, 155–6, 162, 183
see also principal-agent dilemma
standardization, 10, **89–91**, **110–14**, 122, 174, **178–80**, 181, 183, 184, 190
Stirling engine, 21, **22**
storage, 3, 4, 13, 14, 30, 32, 33, 92, 94, 96, 105, 106, 110, 122, **131–9**, 140–9, 159, 182, 187, 188, 191, 192, 209
strategic differentiation, **66**, 87

Telegestore, **128**
Thatcher, Margaret, **10**, 15
Töpfer, Klaus, **15**
transaction costs, **156–7**, 193

unbundling, 1, 93, 95, 107, **110**, 129, 192–3

virtual power plants, 3, 29, **32–4**, 40, 56, 61, 83, 104, 188

Westinghouse, George, **9**, 15
White Certificates, **160**, 212n2

Companies and organizations

Argentus, 3, 105, 109, 119, **173–82**, 190, 194–5
Arthur D. Little, **17**
Arup, **170–1**

Bioenergy village Jühnde, 3, **68–80**, 191, 195–6

Capstone, 51, **58**, 190, 201
co2online, 3, 32, 65, 126–7, 154–5, 160, 162, 163, **164–9**, 193, 196

Daimler, 3, 49, 140, 141, 142, 143, **144–8**, 186, 191, 197
Deutsche Telekom, 37, **40**, 61, 188
Druk White Lotus School, **170–1**

E.ON, 3, 29, 31, 38, **39–40**, 61, 74–5, 107, 118, 124, 125, **126**, 132, **140–1**, **148**, 172, 186, 198
Ecco Solar Group, **85–6**
Echelon, **90**
EnBW Regional, 3, 92, 93, **96–8**, **106–7**, 114, 115, 118, 186, 192, 199
Enel, 90, **128**

GASAG, 3, 25, 31, 80, 84, 144, 159, **161**, 175–6, 200
Greenvironment, 3, 25, **51–9**, 190, 201

Himin Group, **85–6**
Honda, 37

Ingersoll-Rand, 51
International Monetary Fund, 10
Itron, 3, **90**, 105, **107–8**, 110, 111, 112, 113, 114, 115, **119–20**, 122, 123, 189, 202

Landis & Gyr, **90**
LichtBlick, 3, 23, **24–5**, 27, 28, 29, 31, **32–7**, 38, 40, 41, 42, 43, 44, 48, 56, 61, 63, 118, 127, 130, 154, 177, 185, 188, 203

Masdar - Mubadala Development Company, **79–80**
MeteoViva, **182**

Nuru Energy, **62–3**

ODR, 3, 69, 80, 93, **99–104**, 105, 109, 110, 117, 119, 121, 122, 124, 134, 135–6, 139–40, 159, 191, 192, 204

Panasonic, 21, 23, 27

Quiet Revolution, **50–1**

RWE, 31, 37, 59, **125**, 195, 206, 207

Siemens, 3, 38, 39, 40, 80, 90, **93–6**, 101, 119, 124, **125–6**, 130, 159, 186, 189, 191, 205
Stadtwerke Krefeld, 3, 30, 68, 80, **81–4**, 117, 120, 178, 206
Stadtwerke Unna, 3, 80, **81–4**, 104, 119, 121, 127, 148, 207

Tokyo Gas, **27**
Turbec, 51, 53

University of Göttingen, **73**, 77, 195

Vaillant, 21, 37
Vattenfall, 37, **40**, 61, 188, 200
Viessmann, 3, 28, 31, 37, 38, 54, 58, **59–61**, 162, 208
VW, 3, 11, 23, 28, 29, 31, 37, 38, **41–9**, 61, 63, 82, 144, 186, 203, 209

World Bank, 10, **65–6**

Younicos, 3, 12, 15, **16**, 30, 69, 104, 116, **132–9**, 142, 161–2, 185, 188, 209–10